Field Guide for Fire Investigators

National Fire Protection Association

Field Guide for Fire Investigators

Richard L.P. Custer

Editor

ArupFire

JONES AND BARTLETT PUBLISHERS

Sudbury, Massachusetts

BOSTON   TORONTO   LONDON   SINGAPORE

Jones and Bartlett Publishers
40 Tall Pine Drive
Sudbury, MA 01776
978-443-5000
www.jbpub.com

Jones and Bartlett Publishers Canada
6339 Ormindale Way
Mississauga, Ontario L5V 1J2
Canada

Jones and Bartlett Publishers International
Barb House, Barb Mews
London W6 7PA
United Kingdom

ISBN 13: 978-07637-4399-4
ISBN-10: 0-7637-4399-2

6048

National Fire Protection Association
1 Batterymarch Park
Quincy, MA 02169
www.NFPA.org

Jones and Bartlett's books and products are available through most bookstores and online booksellers. To contact Jones and Bartlett Publishers directly, call 800-832-0034, fax 978-443-8000, or visit our website www.jbpub.com.

Substantial discounts on bulk quantities of Jones and Bartlett's publications are available to corporations, professional associations, and other qualified organizations. For details and specific discount information, contact the special sales department at Jones and Bartlett via the above contact information or send an email to specialsales@jbpub.com.

Production Credits
Chief Executive Officer: Clayton E. Jones
Chief Operating Officer: Donald W. Jones, Jr.
President, Higher Education and Professional Publishing: Robert W. Holland, Jr.
V.P., Sales and Marketing: William J. Kane
V.P., Production and Design: Anne Spencer
V.P., Manufacturing and Inventory Control: Therese Connell
Publisher, Public Safety Group: Kimberly Brophy
Acquisition Editor: William Larkin
Editor: Jennifer Reed
Production Editor: Jenny L. McIsaac
Director of Marketing: Alisha Weisman
Interior Design: Anne Spencer
Cover Design: Kristin E. Ohlin
Text Printing and Binding: Replika Press Pvt Ltd.
Cover Printing: Replika Press Pvt Ltd.
Photo Research Manager/Photographer: Kimberly Potvin

Printed in India

11 10 09 08 07   10 9 8 7 6 5 4 3 2

CONTENTS

CONTENTS

Sample Documentation Photographs   2-39

Burn Pattern Identification   2-47

CONTENTS

vii

CONTENTS

PART 3
BUILDING CONSTRUCTION AND SYSTEMS   3-1

CONTENTS

ix

CONTENTS

CONTENTS

PART 4
INFORMATION FOR THE FIRE INVESTIGATOR   4-1

CONTENTS

xiii

CONTENTS

CONTENTS

PREFACE

Over the past 25 years or so, fire investigation has evolved into a more rigorous discipline than ever before. The publication of NFPA 921, *Guide for Fire and Explosion Investigations,* advances in fire science, and the application of information and techniques from other disciplines—such as materials science and failure analysis—have greatly influenced the evolution of fire investigation. As a result of this evolution, the fire investigator needs easy access to a wide variety of information. Because it is rarely practical to carry multiple reference books to the fire scene, this *Field Guide* was created.

The *Field Guide for Fire Investigators* is a portable, convenient reference that contains data, diagrams, tables, forms, contact information, and other useful information for public and private fire investigators to use both in the field and after visiting the fire scene. By compiling essential and practical information from a variety of sources, the *Guide* belongs in every fire investigator's field bag.

The *Field Guide* has four parts. **Part 1, Before Going to the Fire Scene,** contains sections on tools and equipment, safety equipment and practices at the fire scene, management of major investigations, and a customizable directory of contact information. **Part 2, Fire Scene Documentation and Analysis,** is intended for use on the scene and includes sections on scene documentation forms, scene documentation diagrams and drawings, scene documentation photographs, burn pattern identification, and analytical tools for the fire investigator. **Part 3, Building Construction and Systems,** is divided into sections on building construction, electrical systems, and fire protection systems. **Part 4, Information for the Fire Investigator,** is largely composed of tables and includes information on properties of ignition sources, properties of materials, burn injury and overpressure damage, other basic information for the fire investigator, and selected metric information for the fire investigator.

The appendix, Chapter Organization of NFPA 921, compares the table of contents of the 2001 and 2004 editions of NFPA 921 by chapter number and subject. Note that all references to NFPA 921 in this book are based on the 2001 edition, which was current when the *Field Guide* was prepared.

The inside back cover features to-scale rulers in metric and English units so that the investigator can have a convenient size reference when photographing small objects.

Richard L. P. Custer

ABOUT THE EDITOR

Richard L. P. Custer served as the chair of the NFPA Technical Committee on Fire Investigation for the first, second, and third editions of NFPA 921, *Guide for Fire and Explosion Investigations.* He subsequently served as the secretary of NFPA 921 for the fourth edition, published in 2001. He also served on the National Fire/Arson Scene Planning Panel of the U.S. Department of Justice's Technical Working Group on Fire and Arson Scene Evidence. A Certified Fire and Explosion Investigator (CFEI) and a Fellow of the Society of Fire Protection Engineers, Mr. Custer is the United States Technical Director for ArupFire, based in Westborough, Massachusetts.

PART 1

BEFORE GOING TO THE FIRE SCENE

Part 1 of the *Field Guide for Fire Investigators* is intended for use while preparing for the field investigation and includes the following elements:

- Tools and Equipment
- Safety Equipment and Practices at the Fire Scene
- Management of Major Investigations
- Directory of Contact Information

TOOLS AND EQUIPMENT

A wide variety of tools and equipment may be needed when investigating a fire scene. These can range from hand tools such as hammers, pliers, screwdrivers, and wrenches to picks, shovels, and a variety of pry bars. Sample bags, clean paint cans, boxes, tarps, and twine or rope can also be needed depending on the nature of the work anticipated.

When operating from a fixed location like a fire station or fire marshal's office, special vehicles with the tools and equipment already packed on board or pre-packed tool and equipment containers may be useful.

When working a remote scene where train or air travel is called for, transporting tools and equipment may be difficult particularly where security restrictions may be in effect. In such cases, it may be necessary to either pre-ship tools and equipment or make arrangements to have them supplied locally at the destination.

In some cases, heavy equipment such as front-end loaders or cranes are needed. To avoid delays at the scene, a list of local providers of such services should be prepared in advance of future need and arrangements made in advance for acquiring and paying for their services.

SAFETY EQUIPMENT AND PRACTICES AT THE FIRE SCENE

Fire scenes are inherently dangerous even after the fire itself has been extinguished. The scene might be dark, obscuring potential hazards including trip and fall hazards (such as holes, debris, exposed wiring, and structural members), broken glass, exposed nails, jagged metal, and the like.

The list of personal safety equipment in Figure 1.1 is based on Chapter 12 of NFPA 921, *Guide for Fire and Explosion Investigations.* The reader is also referred to *Safety and Health Guidelines for Fire and Explosion Investigators,* published in 2002 by the International Fire Service Training Association (IFSTA) at Oklahoma State University.

Each person on the fire scene should be equipped with appropriate safety equipment, as required. A complement of basic tools should also be available. The tools and equipment listed below may not be needed on every scene, but in planning the investigation, the investigator should know where to obtain these tools and equipment if the investigator does not carry them.

Personal Safety Equipment

Eye protection
Flashlight
Gloves
Helmet or hard hat

Respiratory protection
(type depending on
exposure)
Safety boots or shoes

Turnout gear or
coveralls

Tools and Equipment

Absorption material
Axe
Broom
Camera and film
(See NFPA 921
13.2.2.1 and
13.2.2.2 for
recommendations.)
Claw hammer
Directional compass
Evidence-collecting
container (See
Section 14.5 of
NFPA 921 for
recommendations.)
Evidence labels
(sticky)

Hand towels
Hatchet
Hydrocarbon detector
Ladder
Lighting
Magnet
Marking pens
Paint brushes
Paper towels/wiping
cloths
Pen knife
Pliers/wire cutters
Pry bar
Rake
Rope
Rulers
Saw

Screwdrivers
(multiple types)
Shovel
Sieve
Soap and hand cleaner
Styrofoam cups
Tape measure
Tape recorder
Tongs
Tweezers
Twine
Voltmeter/ohmmeter
Water
Writing/drawing
equipment

Figure 1.1 Equipment and facilities.
Source: NFPA 921, 12.4.1.

MANAGEMENT OF MAJOR INVESTIGATIONS

A major fire does not necessarily have to be a large fire physically, financially, or in terms of death or injury. While major fires may have one or more of the above characteristics, what they have in common is that they are complex incidents where the primary goals are to preserve evidence and protect the interests of the interested parties whether or not they can be identified at the time of the scene investigation. The following are suggested essential elements for managing a major fire:

- Allow interested parties to participate
- Allow evidence to be examined in undisturbed condition to the extent possible
- Establish a written agreement between parties covering, where appropriate:
  - Fire scene security (control and access to the site)
  - Sharing of information (factual material such as photos, building and fire protection plans, specifications, interviews, etc.)
  - Protocol for scene examination, documentation, debris removal
  - Joint custody and joint examination of evidence
  - Notification of parties and development of protocols before destructive testing
  - Acquisition and processing of non-proprietary data and information
  - Payment of costs incurred (fencing, lighting, sanitation, heavy equipment, surveying, photography, etc.)
- Assign specific person (team leader) to represent each party
- Use Team Leader Committee and regular meetings for coordination of activities
- Establish safety protocols (protective gear, required HAZMAT certification, evacuation signals, etc.)
- Conduct primary witness interviews with interested parties represented where possible

Some written agreement or memo of understanding should be prepared whenever there are multiple "interested parties" working at a fire scene. If possible it is best to get both public and private parties to be in agreement regarding procedures for the processing of the scene and evidence collection and testing. While this may not always be possible, once the scene is released, the private parties should work out an agreement covering details such as those listed above. See Figure 1.2 for a sample memorandum of understanding.

Any fire investigation requires the investigator to have a plan on how to proceed. Nowhere is the need for a plan more important than when managing a major fire investigation. Figure 1.3 illustrates a sample flow chart for a fire investigation. How the fire investigator implements such a flow chart is essentially the fire investigation plan.

Details of this process are presented in the Management of Major Investigations in Chapter 24 of NFPA 921.

BEFORE GOING TO THE FIRE SCENE
Management of Major Investigations

This Memorandum of Understanding relates to the investigation of the fire that occurred on July 1, 1998, at the Tall Building and Storage Facility, 1007 Main Ave., Any City, State, USA. It recognizes that a number of independent investigations are being conducted simultaneously and coincidentally and all with a common goal — to determine the origin and cause of the fire. All interested parties recognize that cooperation with one another will be beneficial to each party and will produce an efficient, quality outcome.

The parties agree to the following:

An origin and cause investigation is being conducted.

The investigation is being conducted by the Yourtown Fire Department, The Federal Fire Investigations, Payall Insurance Company, Any Storage Company, and the Tall Building Company.

All investigation procedures and the physical collection of the evidence will be coordinated through regular meetings. The evidence will be collected and stored in a location where access is monitored. No testing or examination of the evidence will be conducted until all parties are notified.

All requests for data of a nonproprietary nature from the Tall Building Company or tenants will be processed through their identified representatives. Nonproprietary information provided by any party will be shared by all parties if requested.

All releases of information regarding the origin and cause of the fire will be coordinated through the Yourtown Fire Department, and no predisclosure of information will be made by any party.

The protocol recognizes that to remove material or conduct testing will require the permission of the Yourtown Fire Department and the undersigned parties. The request should be in writing; however, verbal agreement followed by written request and approval of the parties will be acceptable when time frames are short.

Testing and examination protocol of materials associated with this investigation are as follows:

(1) All parties agree as to who will perform each examination and each test.

(2) All parties agree to allow any other party to observe each test.

(3) All parties agree to return any material remaining after each test to the storage facility.

Attached is an investigation flow chart to provide guidance for the general scope of the investigation.

Figure 1.2 Sample memorandum of understanding.
Source: NFPA 921, Figure 24.3(a).

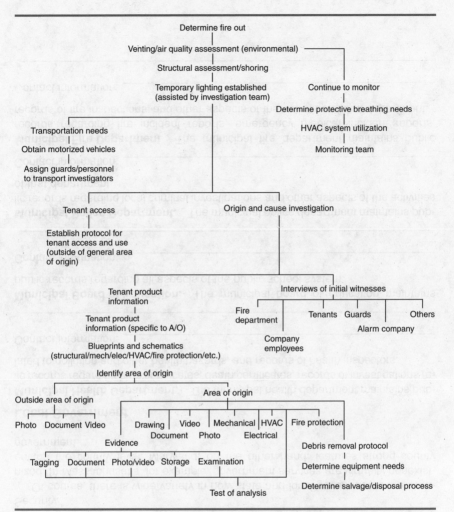

Figure 1.3 Investigation flow chart.
Source: NFPA 921, Figure 24.3(b).

DIRECTORY OF CONTACT INFORMATION

In the course of field work, the fire investigator relies on information available from a number of local, state, federal, and non-profit organizations. The list and descriptions of contact organizations that follow are based on NFPA 921, Chapter 11. The fire investigator is encouraged to note addresses and phone numbers of local and state organizations in the blank space provided. Contact information for United States federal and nonprofit organizations is current as of May 2003 and reflects the reorganization of the federal government to add the Department of Homeland Security.

Of course, there is wide variety in how state and local governments are organized. In Massachusetts, for example, government services are virtually nonexistent on the county level. Maryland, on the other hand, features strong county government.

Local Government

Municipal Health Department. The municipal health department maintains public records regarding birth certificates, death certificates, records of investigations related to pollution and other health hazards, and records of health inspectors.

Contact information: _____

_____

Municipal Board of Education. The municipal board of education maintains public records regarding all aspects of the public school system.

Contact information: _____

_____

Municipal Police Department. The municipal police department maintains public records regarding local criminal investigations and other aspects of the activities of that department.

Contact information: _____

Municipal Fire Department. The municipal fire department maintains public records regarding fire incident reports, emergency medical incident reports, records of fire inspections, and other aspects of the activities of that department.

Contact information: _____

_____

Other Municipal Agencies. Many other offices, departments, and agencies typically exist at the municipal level of government. The fire investigator may encounter different governmental structuring in each municipality. As such, the fire investigator may need to solicit information from these additional sources.

Contact information: _____

_____

County Government

County Recorder. The county recorder's office maintains public records regarding documents relating to real estate transactions, mortgages, certificates of marriage and marriage contracts, divorces, wills admitted to probate, official bonds, notices of mechanics' liens, birth certificates, death certificates, papers in connection with bankruptcy, and other such writings as are required or permitted by law.

Contact information: _____

_____

County Clerk. The county clerk maintains public records regarding naturalization records, civil litigation records, probate records, criminal litigation records, and records of general county business.

Contact information: _____

_____

County Assessor. The county assessor maintains public records such as plats or maps of real property in the county, which include dimensions, addresses, owners, and taxable value.

Contact information: _____

_____

County Treasurer. The county treasurer maintains public records regarding names and addresses of property owners, names and addresses of taxpayers, legal descriptions of property, amounts of taxes paid or owed on real and personal property, and all county fiscal transactions.

Contact information: _____

_____

BEFORE GOING TO THE FIRE SCENE
Directory of Contact Information

County Coroner/Medical Examiner. The county coroner/medical examiner maintains public records regarding the names or descriptions of the deceased, dates of inquests, property found on the deceased, causes and manners of death, and documents regarding the disposition of the deceased.

Contact information: _____

_____

County Sheriff's Department. The county sheriff's department maintains public records regarding county criminal investigations and other aspects of the activities of that department.

Contact information: _____

_____

Other County Agencies. Many other offices, departments, and agencies typically exist at the county level of government. The fire investigator may encounter different governmental structuring in each county. As such, the fire investigator may need to solicit information from these additional sources.

Contact information: _____

_____

State Government

Secretary of State. The secretary of state maintains public records regarding charters and annual reports of corporations, annexations, and charter ordinances of towns, villages, and cities; trade names and trademarks registration; notary public records; and Uniform Commercial Code (UCC) statements.

Contact information: _____

_____

State Treasurer. The state treasurer maintains public records regarding all state fiscal transactions.

Contact information: _____

_____

State Department of Vital Statistics. The state department of vital statistics maintains public records regarding births, deaths, and marriages.

Contact information: _____

_____

State Department of Revenue.   The state department of revenue maintains public records regarding individual state tax returns; corporate state tax returns; and past, present, and pending investigations.

Contact information: _____

_____

State Department of Regulation.   The state department of regulation maintains public records regarding names of professional occupation license holders and their backgrounds; results of licensing examinations; consumer complaints; past, present, or pending investigations; and the annual reports of charitable organizations.

Contact information: _____

_____

State Department of Transportation.   The state department of transportation maintains public records regarding highway construction and improvement projects, motor vehicle accident information, motor vehicle registrations, and driver's license testing and registration.

Contact information: _____

_____

State Department of Natural Resources.   The state department of natural resources maintains public records regarding fish and game regulations, fishing and hunting license data, recreational vehicles license data, waste disposal regulation, and environmental protection regulation.

Contact information: _____

_____

State Insurance Commissioner's Office.   The state insurance commissioner's office maintains public records regarding insurance companies licensed to transact business in the state; licensed insurance agents; consumer complaints; and records of past, present, or pending investigations.

Contact information: _____

_____

State Police.   The state police maintain public records regarding state criminal investigations and other aspects of the activities of that agency.

Contact information: _____

_____

BEFORE GOING TO THE FIRE SCENE
Directory of Contact Information

BEFORE GOING TO THE FIRE SCENE
Directory of Contact Information

State Fire Marshal's Office. The state fire marshal's office maintains public records regarding fire inspection and prevention activities, fire incident databases, and fire investigation activities.

Contact information: _____

_____

Other State Agencies. Many other offices, departments, and agencies typically exist at the state level of government. The fire investigator may encounter different government structuring in each state. As such, the fire investigator may need to solicit information from these additional sources.

Contact information: _____

_____

U.S. Federal Government

The names and organizations of many agencies in the U.S. federal government changed with the establishment of the new Department of Homeland Security. The following contact information for the U.S. federal government is current as of May 2003.

U.S. Department of Agriculture

- The former **Consumer & Marketing Service,** now Agricultural Marketing Service (AMS), P.O. Box 96456, Washington, D.C. 20250; 202-720-8999; *www.ams.usda.gov*
- **Grain Inspection, Packers & Stockyards Administration,** P.O. Box 96454, Washington, D.C. 20090-6454; 202-720-5091
- **Dairy Product Inspection,** *www.ams.usda.gov/admin/amsorg.htm*
- The former **Meat/Poultry, Egg Poultry Inspection Service,** now Food Safety & Inspection Service (FSIS), USDA, Washington, D.C. 20250; *www.fsis.usda.gov*
- **U.S. Forest Service,** Mail Stop 1125, 1400 Independence Avenue, S.W., Washington, D.C. 20090-1125; 202-205-1706; *www.fs.fed.us*

U.S. Department of Commerce

- **Headquarters,** 1401 Constitution Ave, N.W., Washington, D.C. 20230; 202-482-2000; *www.commerce.gov*
- The former **Commercial Intelligence Division Office,** now the International Trade Administration (ITA), Commercial Service Division, 14th & Constitution Avenue, N.W., Washington, D.C. 20230; 202-482-4204; *www.ita.doc.gov*

- **National Marine Fisheries Service,** 1315 East-West Hwy, SSMC3, Silver Spring, MD 20910; 301-713-2334; *www.nmfs.gov*

- **U.S. Patent Office,** General Information Services Division, Crystal Plaza 3, Room 2C02, P.O. Box 1450, Alexandria, VA 22313-1450; 800-786-9199 or 703-308-4357; *www.uspto.gov*

- The **Trade Mission Division,** Department of Commerce, 1401 Constitution Ave, N.W., Washington, D.C. 20230; 800-872-8723 or 202-482-0003; *www.ita.doc.gov*

- **National Climatic Data Center.** Federal Building, 151 Patton Avenue, Asheville NC 28801-5001; 828-271-4800; *www.ncdc.noaa.gov/oa/ncdc.html*

- **National Oceanic and Atmospheric Administration,** Department of Commerce, 14th Street & Constitution Avenue, N.W., Room 6217, Washington, D.C. 20230; 202-482-6090; *www.noaa.gov*

U.S. Department of Defense. Directorate for Public Inquiry and Analysis, Room 3A750–The Pentagon, 1400 Defense Pentagon, Washington, D.C. 20301-1400; 703-545-6700; *http://www.defenselink.mil/*

U.S. Department of Energy. 1000 Independence Ave., S.W., Washington, D.C. 20585; 800-dial-DOE; *www.energy.gov*

U.S. Department of Health and Human Services

- **Headquarters,** 200 Independence Ave, S.W., Washington, D.C. 20201; toll free 877-696-6775 or 202-619-0257; *http://www.os.dhhs.gov/*

- **Social Security Administration,** 6401 Security Blvd., Baltimore, MD 21235-0001; 800-772-1213; *www.socialsecurity.gov*

- **U.S. Department of Housing and Urban Development.** Headquarters, 451 7th Street S.W., Washington, D.C. 20410; 202-708-1112; TTY: (202) 708-1455; *http://www.hud.gov/*

Department of the Interior

- **Headquarters,** 1849 C Street N.W., Washington, D.C. 20240; 202-208-3100; *http://www.doi.gov*

- **Bureau of Indian Affairs,** 1849 C Street N.W., Washington, D.C. 20240; 202-208-3710; *www.doi.gov/bureau-indian-affairs.html*

- **National Park Service,** 1849 C Street N.W., Washington, D.C. 20240; 202-208-6843; *www.nps.gov*

U.S. Department of Justice

- **Headquarters,** 950 Pennsylvania Avenue, N.W., Washington, D.C. 20530-0001; 202-353-1555; *www.usdoj.gov*

- **The Civil Rights Division,** Department of Justice, 950 Pennsylvania Ave., N.W., Washington, D.C. 20530; 202-514-2151; *www.usdoj.gov/crt/crt-home.html*

- **The Criminal Division,** Department of Justice, 950 Pennsylvania Ave., N.W., Washington, D.C. 20530; 202-514-2601; *http://www.usdoj.gov/criminal*

- **Drug Enforcement Administration,** Department of Justice, Mailstop: AXS, 2401 Jefferson Davis Highway, Alexandria, VA 22301; 800-882-9539 or 202-307-8000; *http://www.usdoj.gov/dea*

- **Federal Bureau of Investigation,** Department of Justice, 935 Pennsylvania Avenue, N.W., Washington, D.C. 20535; 202-324-3000; *www.fbi.gov*

- **Immigration and Naturalization Service** (formerly part of the U.S. Department of Justice and now a bureau of the Department of Homeland Security), 4420 N. Fairfax Drive, Arlington, VA 22203; 800-375-5283; *www.immigration.gov*

U.S. Department of Labor

- **Headquarters,** Frances Perkins Building, 200 Constitution Avenue, N.W., Washington, D.C. 20210; 866-4-USA-DOL; TTY: 1-877-889-5627; *www.dol.gov*

- **Employment Standards Administration,** Frances Perkins Building, 200 Constitution Avenue, N.W., Washington, D.C. 20210; 866-4-USA-DOL; TTY: 1-877-889-5627; *www.dol.gov/esa*

U.S. Department of State, 2201 C Street, N.W., Washington, D.C. 20520; 202-647-4000; *www.state.gov*

U.S. Department of Transportation

- **Headquarters,** 400 7th Street. S.W., Washington, D.C. 20590; 202-366-4000; *www.dot.gov*

- **United States Coast Guard,** 2100 Second Street, S.W., Washington, D.C. 20593; 202-267-2229; *www.uscg.mil/USCG.shtm*

U.S. Department of the Treasury

- **Headquarters,** 1500 Pennsylvania Avenue, N.W., Room 2134, Washington, D.C. 20220, 202-622-2000; *www.ustreas.gov*

- **Bureau of Alcohol, Tobacco, and Firearms,** 650 Massachusetts Avenue, N.W., Room 8290, Washington, D.C. 20226; 202-927-7777; *www.atf.treas.gov*

- **U.S. Customs Service** (formerly within the Department of the Treasury and now part of the Department of Homeland Security), Department of Homeland Security, 1300 Pennsylvania Ave, N.W., Washington, D.C. 20229; 202-354-1000; *www.customs.ustreas.gov*

- **Internal Revenue Service,** 500 N. Capitol St., N.W., Washington, D.C. 20221; 202-874-6748; *www.irs.ustreas.gov*

- **U.S. Secret Service,** 950 H Street, N.W., Suite 8400, Washington, D.C. 20223; 202-406-5708; *www.secretservice.gov*

U.S. Fire Administration, 16825 S. Seton Ave., Emmitsburg, MD 21727, 301-447-1000*; www.usfa.fema.gov*

U.S. Postal Service, Postmaster General, 475 L'Enfant Plaza Place, S.W., Washington, D.C. 20260; 800-275-8777, *www.usps.com*

Not-for-Profit Organizations

- **American Society for Testing and Materials (ASTM),** 100 Barr Harbor Drive, P.O. Box C700, W. Conshohocken, PA 19428-2959; 610-832-9585; *www.astm.org*

- **International Association of Arson Investigators (IAAI),** 12770 Boenker Road, Bridgeton, MO 63044; 314-739-4224; *www.firearson.com*

- **National Association of Fire Investigators (NAFI),** 857 Tallevast Road, Sarasota, FL 34243; 877-506-NAFI; *www.nafi.org*

- **National Fire Protection Association (NFPA),** 1 Batterymarch Park, Quincy, MA 02269; 617-770-3000; *www.nfpa.org*

- **Society of Fire Protection Engineers (SFPE),** 7315 Wisconsin Avenue, Suite 1225 W, Bethesda, MD 20814; 301-718-2910; *www.sfpe.org*

PART 2

FIRE SCENE DOCUMENTATION AND ANALYSIS

Part 2 of the *Field Guide for Fire Investigators* is intended for use on the investigation scene and includes the following elements:

- Scene Documentation Forms
- Scene Documentation Diagrams and Drawings
- Scene Documentation Photographs
- Burn Pattern Identification
- Analytical Tools for the Fire Investigator

SCENE DOCUMENTATION FORMS

When documenting a fire scene, it is frequently helpful to employ forms or lists to help organize data for future use and as a reminder of items that may be needed for complete documentation. It is important, however, not to become overly reliant upon these aids to the extent that items are missed because they are not "on the list." On the other hand, it may not be possible or necessary to address everything on a given list or form. The experienced investigator should use judgment in the application of these aids, adding and subtracting as necessary to provide documentation. Remember…once the scene is gone, only the documentation remains for future study and analysis.

Figures 2.1 through 2.6 are forms for the fire investigator's use in documenting the scene. Note that Figure 2.5 is from the 2001 edition of NFPA 921, *Guide for Fire and Explosion Investigations;* the human figure from that chart is included in the form (Figure 2.6) that has been proposed for the 2004 edition of NFPA 921.

Evidence Tags and Chain of Custody Forms

Marking or tagging of evidence is important for future identification of the items with regard to the details of their collection and the persons or organization collecting them. It is helpful to have pre-printed tags or labels and a means of attaching them to the evidence or artifacts without causing damage. A form or log of the evidence collected is also valuable. A sample evidence tag and collection form is provided below. When there are multiple interests represented at a fire scene, a number of them may be interested in the same item of evidence, in which case each party should put their own tag on the sample. See Figure 2.7.

FIRE SCENE DOCUMENTATION AND ANALYSIS
Scene Documentation Forms

STRUCTURE FIRE

Agency			Case number	

TYPE OF OCCUPANCY

Residential		Single family	Multifamily	Commercial		Governmental
Church	School		Other:			
Estimated age:		Height (stories):		Length:		Width:

PROPERTY STATUS

Occupied at time of fire? ❏ Y ❏ N	Unoccupied at time of fire? ❏ Y ❏ N	Vacant at time of fire? ❏ Y ❏ N
Name of person last in structure prior to fire:	Time and date in structure	Exited via which door/egress:
Remarks:		

BUILDING CONSTRUCTION

Foundation Type	Basement		Crawl Space		Slab			Other:	
Material	Masonry		Concrete		Stone			Other:	
Exterior Covering	Wood	Brick/Stone	Vinyl		Asphalt	Metal	Concrete		Other:
Roof	Asphalt		Wood	Tile		Metal		Other:	
Type of Construction	Wood Frame	Balloon	Heavy Timber	Ordinary		Fire Resistive	Non combustible		Other:

ALARM/PROTECTION/SECURITY

Sprinklers ❏ Y ❏ N	Standpipes ❏ Y ❏ N	Security Camera(s) ❏ Y ❏ N
Smoke Detectors ❏ Y ❏ N	Hardwired ❏ Y ❏ N	Battery ❏ Y ❏ N
Were Batteries in place ❏ Y ❏ N	Location(s):	
Hidden Keys ❏ Y ❏ N Where:	Security Bars Windows? ❏ Y ❏ N	Doors? ❏ Y ❏ N
Remarks:		

CONDITIONS DOORS/WINDOWS

Doors	Locked		Unlocked but closed	Open	
	Forced Entry ❏ Y ❏ N		Who forced if known?		
Windows	Secure	Unlocked but closed	Open		Broken
	Broken by first responders ❏ Y ❏ N		Remarks		

Figure 2.1 Structure fire notes form.

FIRE DEPARTMENT OBSERVATIONS

Name of first on scene:	Department

General Observations:

Obstacles to extinguishment?	First-In Report Attached ☐ Y ☐ N

UTILITIES

Electric	On	Off	None		Overhead	Underground
	Company			Contact		Telephone
Gas/Fuel	On	Off	None		Natural	LP	Oil
	Company			Contact		Telephone
Water	Company			Contact		Telephone
Telephone	Company			Contact		Telephone
Other	Company			Contact		Telephone

COMMENTS

Figure 2.1 *(continued)*

FIRE SCENE DOCUMENTATION AND ANALYSIS
Scene Documentation Forms

FIRE INCIDENT
FIELD NOTES

Agency:	File No.:

INCIDENT

Location/Address						
Property Description	Structure	Residential	Commercial	Vehicle	Wildland	Other
Other Relevant Info						

WEATHER CONDITIONS

Indicate Relevant Weather Information	Visibility	Rel. Humidity	GPS		Elevation	Lightning
	Temperature	Wind Direction		Wind Speed	Precipitation	

OWNER

Name		DOB
d/b/a (if applicable)		
Address		
Telephone	Home	Business	Cellular

OCCUPANT

Name		DOB
d/b/a (if applicable)		
Permanent Address		
Temporary Address		
Telephone	Home	Business	Cellular

DISCOVERED BY

Incident discovered by	Name/Address	DOB
Telephone	Home	Business	Cellular

Figure 2.2 Fire incident field notes form.

Fire Incident Field Notes Continued
File No.:

REPORTED BY

Incident reported by	Name/Address		DOB
Telephone	Home	Business	Cellular

INVESTIGATION INITIATION

Request date and time	Date of request	Time of request
Investigation requested by	Agency name	Contact person/telephone no.
Request received by	Agency name	Contact person/telephone no.

SCENE INFORMATION

Arrival information	Date		Time		Comments		
Scene secured	No	Yes	Securing Agency		Manner of security		
Authority to enter	Contemporaneous to exigency		Consent		Warrant		
			Written	Verbal	Admin.	Crim.	Other
Departure information	Date		Time		Comments		

OTHER AGENCIES INVOLVED

Primary fire department	Dept. or Agency Name	Incident No.	Contact Person/Phone
Secondary fire department(s)			
Law enforcement			
Private investigators			

ADDITIONAL REMARKS

Figure 2.2 *(continued)*

FIRE SCENE DOCUMENTATION AND ANALYSIS
Scene Documentation Forms

FIRE SCENE DOCUMENTATION AND ANALYSIS
Scene Documentation Forms

VEHICLE INSPECTION FIELD NOTES

Job # _____ Claim # _____ Date of Loss _____
Insured _____ Date of Assignment _____
Address (City , State) _____ Date of Receipt _____
Loss Location _____ Date of Inspection _____
_____ Insp Location _____
Stolen? ❑ Yes ❑ No Recovered By _____ at _____ on _____
Police Report _____ Fire Report _____
# of Keys _____ Alarm System? ❑ Yes ❑ No Alarm Type _____
Hidden Keys? ❑ Yes ❑ No Location _____

VEHICLE
Make _____ Model _____ Year _____
VIN _____ Odometer _____

EXTERIOR

Tires	Tire Type	Wheel Type	Tire Tread Depth	Lugs	Missing
LF					
LR					
RR					
RF					
SP					

Doors	Glass Y/N	Window UP/DOWN	(un)Locked	Open/Closed	Prior Damage
LF					
LR					
RR					
RF					

Body Panels	Construction	Condition	Prior Damage
F Bumper			
Grill			
LF Fender			
LR Quarter			
R Bumper			
RR Quarter			
RF Fender			
Hood			
Roof			
Trunk			

UNDER HOOD	Intact	Missing	Parts Missing	Condition
Engine				
Battery				
Belts & Hoses				
Wiring				
Accessories				

FLUIDS	Level	Condition	Sample Taken
Oil			
Transmission			
Radiator			
Pwr Steer			
Brake			
Clutch			

(Vehicle Inspection Field Notes 1 of 2)

Figure 2.3 Vehicle fire field notes form.
Source: NFPA 921, 2001 edition, Figure 22.7.3.

Job # _____

INTERIOR	Intact	Missing	Parts Missing	Condition
Dash Pod				
Glove Box				
Strg Column				
Ignition				
Front Seat				
Rear Seat				
Rear Deck				
			Make/Model	
Stereo				
Speakers				
Accessories				
FLOOR			Sample Taken	
LF				
LR				
RR				
RL				

PERSONAL EFFECTS IN THE INTERIOR

TRUNK OR CARGO AREA

AFTERMARKET ITEMS NOT PREVIOUSLY DESCRIBED

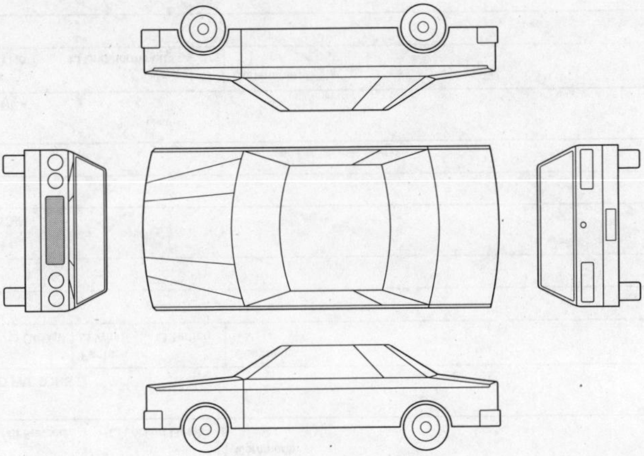

ATS 851B, 8/97

(Vehicle Inspection Field Notes 2 of 2)

Figure 2.3 *(continued)*

FIRE SCENE DOCUMENTATION AND ANALYSIS
Scene Documentation Forms

FIRE SCENE DOCUMENTATION AND ANALYSIS
Scene Documentation Forms

WILDFIRE NOTES

Agency	File number

PROPERTY DESCRIPTION

Fire damage ☐ Less than acre _____ No. Acres	Other Properties Involved
Security ☐ Open ☐ Fenced ☐ Locked Gates	Comments

FIRE SPREAD FACTORS

Type Fire ☐ Ground ☐ Crown	Factors ☐ Wind ☐ Terrain	Comments

AREA OF ORIGIN

PEOPLE IN AREA

At time of fire ☐ Yes ☐ No ☐ Undetermined	Comments

IGNITION SEQUENCE

Heat of ignition			
Material ignited			
Ignition factor			
If equipment involved	Make	Model	Serial No.
Comments			

Figure 2.4 Wildfire notes form.

BODY DIAGRAM

Indicate parts of body injured:
   [ ] None   [ ] Blisters (red marker)   [ ] Burns (black marker)

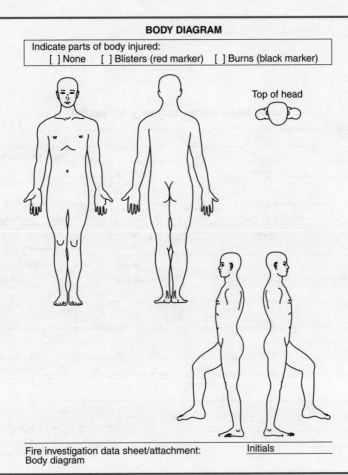

Top of head

Fire investigation data sheet/attachment: Initials _____
Body diagram

Figure 2.5 Example of a chart that can be used to diagram injuries.
Source: NFPA 921, 2001 edition, Figure 20.6.10.

FIRE SCENE DOCUMENTATION AND ANALYSIS
Scene Documentation Forms

CASUALTY VICTIM NOTES

Agency	File number

DESCRIPTION

Name		Address				Phone No.	
Race	Sex	Date of birth	Height	Weight	Hair	Eyes	Other
Describe clothing							

TYPE OF INJURY

❑ Minor　　❑ Moderate　　❑ Severe　　❑ Fatal	Describe Injury

CIRCUMSTANCES

Who found victim?	Where?
Victim's activity just prior to and at time of ignition	
Victim's activity after time of ignition	

CASUALTY TREATMENT

❑ Treated at scene by?		
Sent to	Via	Treated by
Remarks		

FATALITIES

Body position and location		
Body removed to	Body removed by	Authority to move body given by
Medical examiner/coroner	Address	Phone no.
Cause of death		
Autopsy by	Address	Phone no.
Date of autopsy　　Case no.	Blood test ❑ YES ❑ NO	X-Rays ❑ YES ❑ NO　　Reports in possession

NEXT OF KIN

Name	Relationship	Address & phone
Notified by (how, date & time)		

Figure 2.6 Casualty victim notes form.

REMARKS

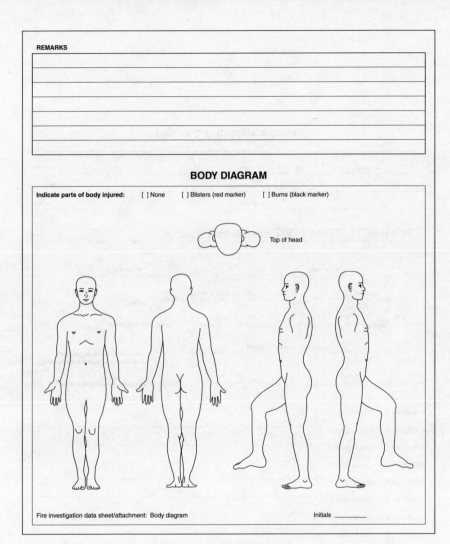

BODY DIAGRAM

Indicate parts of body injured: [ ] None   [ ] Blisters (red marker)   [ ] Burns (black marker)

Top of head

Fire investigation data sheet/attachment: Body diagram   Initials _____

Figure 2.6 *(continued)*

FIRE SCENE DOCUMENTATION AND ANALYSIS
Scene Documentation Forms

EVIDENCE FORM

Date of Incidents _____/_____/_____   Storage Location: _____   Case No. _____

Item No.	DESCRIPTION	LOCATION		
_____	_____	_____	❑ DESTROYED	❑ RELEASED
_____	_____	_____	❑ DESTROYED	❑ RELEASED
_____	_____	_____	❑ DESTROYED	❑ RELEASED
_____	_____	_____	❑ DESTROYED	❑ RELEASED
_____	_____	_____	❑ DESTROYED	❑ RELEASED
_____	_____	_____	❑ DESTROYED	❑ RELEASED
_____	_____	_____	❑ DESTROYED	❑ RELEASED
_____	_____	_____	❑ DESTROYED	❑ RELEASED
_____	_____	_____	❑ DESTROYED	❑ RELEASED
_____	_____	_____	❑ DESTROYED	❑ RELEASED

How was evidence received?   Date received: _____/_____/_____   Date stored: _____/_____/_____

Removed from scene by investigator.
Received by investigator from: _____
NAME, COMPANY, OR DEPT.

Received via: ❑ UPS   ❑ FEDEX   ❑ AIRBORNE   ❑ USMAIL   ❑ IN PERSON

Freight _____   **Other** _____
Name of company   Describe

_____   _____
Received by   Case investigator

LOCATION EVIDENCE REMOVED

Owner _____   Address 2 _____

Company _____   City _____

Address 1 _____   State _____ Zip _____ Phone _____

Figure 2.7 Evidence form.

INTERNAL EXAMINATION

Investigator	Date Pulled	Date Examined	Date Returned

EVIDENCE DESTRUCTION

Authorized by _____ Date _____

Investigator's authorization _____ Date _____

Destroyed by _____ Date _____

EVIDENCE RELEASE

Signature of person receiving evidence _____

Person receiving evidence (please print) _____ Date _____

Company name _____

Address _____

City _____ State _____ Zip code _____

Authorized by _____ Date _____

Investigator's authorization _____ Date _____

Released VIA _____

Remarks _____
_____
_____
_____
_____

EXAMINATION BY OTHERS

Name _____ Date of examination _____

Company _____

Address _____

City _____ State _____ Zip code _____

Phone _____

Authorized by _____

Investigator's authorization _____ Date _____

Name _____ Date of examination _____

Company _____

Address _____

City _____ State _____ Zip code _____

Phone _____

Authorized by _____

Investigator's authorization _____ Date _____

Name _____ Date of examination _____

Company _____

Address _____

City _____ State _____ Zip code _____

Phone _____

Authorized by _____

Investigator's authorization _____ Date _____

Figure 2.7 *(continued)*

FIRE SCENE DOCUMENTATION AND ANALYSIS
Scene Documentation Forms

Once evidence is collected, some means should be established to keep track of the transfer of custody of the evidence and the examination or testing that is performed. An example of a "chain of custody" form appears as Figure 2.8.

Scene Documentation Diagrams and Drawings

Sketches and diagrams are one of the most useful methods for documentation of the fire scene. Diagrams can be used to document information from a large area such as the locations of fire apparatus on the scene and the locations of fire fighting activities. They may also be used to identify the locations of witness or evi-

---

Crime Scene Search Evidence Report

Name of subject _____

Offense _____

Date of incident _____ Time _____ a.m. p.m.

Search officer _____

Evidence description _____

_____

Location _____

_____

Chain of Possession

Received from _____

By _____

Date _____ Time _____ a.m. p.m.

Received from _____

By _____

Date _____ Time _____ a.m. p.m.

Received from _____

By _____

Date _____ Time _____ a.m. p.m.

Received from _____

By _____

Date _____ Time _____ a.m. p.m.

Figure 2.8 Chain of custody form.
Source: NFPA 921, 2001 edition, Figure 14.9.

dence samples, show relationships between objects at a possible point of fire origin, or present overall fire damage patterns in an easy to understand format. The amount of information presented on a diagram or sketch can be greatly increased by the use of standard symbols such as those provided in this section. When possible, measurements should be made of building dimensions, doors, windows, and interior partitions. Since photocopies and enlargements are frequently made of fire scene diagrams, it is very useful to include a graphic scale on the drawing that will be enlarged or reduced along with the drawing.

Table 2.1 lists design and construction diagrams that may be available for the fire investigator's review.

Table 2.1 Design and Construction Drawings That May Be Available

Type	Information	Discipline
Topographical	Shows the varying grade of the land	Surveyor
Site plan	Shows the structure on the property with sewer, water, electrical distributions to the structure	Civil engineer
Floor plan	Shows the walls and rooms of structure as if you were looking down on it	Architect
Plumbing	Layout and size of piping for fresh and waste water	Mechanical engineer
Electrical	Size and arrangement of service entrance, switches and outlets, fixed electrical appliances	Electrical engineer
Mechanical	HVAC system	Mechanical engineer
Sprinkler/fire alarm	Self-explanatory	Fire protection engineer
Structural	Frame of building	Structural engineer
Elevations	Shows interior/exterior walls	Architect
Cross section	Shows what the inside of components look like if cut through	Architect
Details	Show close-ups of complex areas	All disciplines

Source: NFPA 921, 2001 edition, Table 13.4.6.

FIRE SCENE DOCUMENTATION AND ANALYSIS
Scene Documentation Forms

Sketching the fire scene is a primary task of the fire investigator. Figures 2.9 and 2.10 illustrate typical site plans and site sketches, respectively.

Especially in major fires, the sketches are likely to be re-rendered as formal diagrams, with or without the use of computer-assisted drawing programs. The use

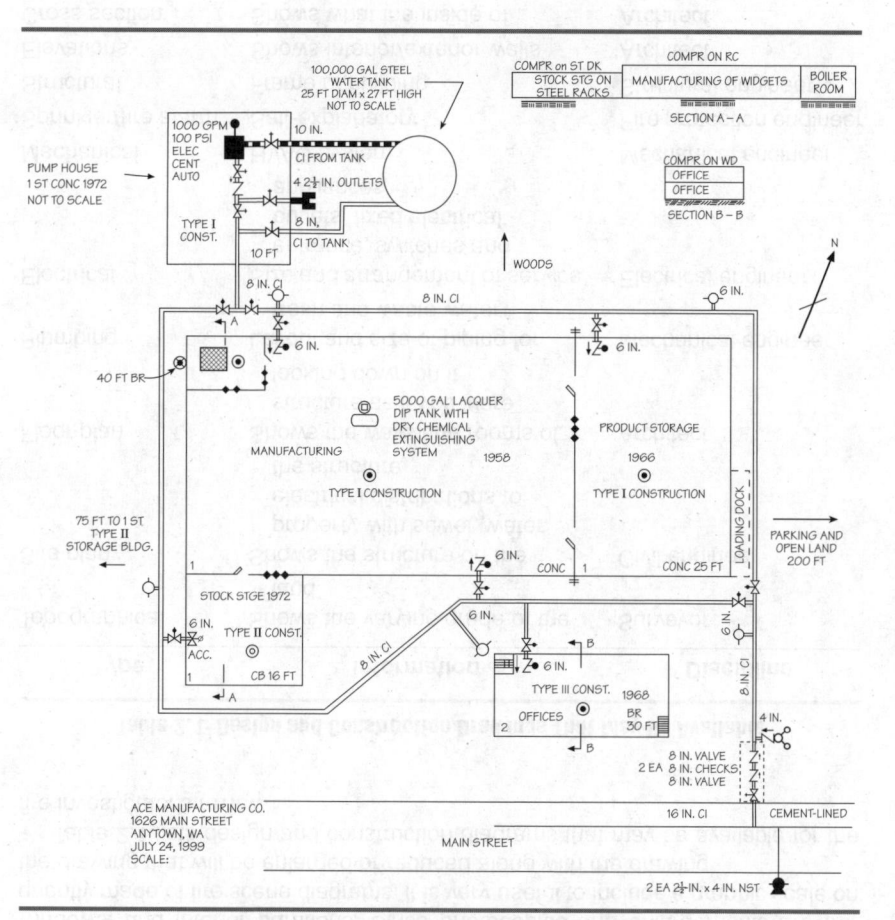

Figure 2.9 Typical site sketch showing fire protection facilities and equipment.

Source: *Fire Protection Handbook,* 19th edition, Figure 3.15.1.

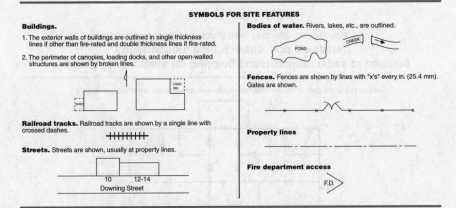

Figure 2.10 Symbols for site features.
Source: *Fire Protection Handbook,* 19th edition, Figure 3.15.2.

of standardized symbols in diagrams and drawings contributes to the professionalism of the fire investigator's work product. NFPA 170, *Standard for Fire Safety Symbols,* provides uniform symbols. Figures 2.11 through 2.29 illustrate standardized symbols of particular use for the fire investigator in documenting a fire scene.

The symbols in Figures 2.27–2.29, while intended for prefire planning, can also be useful for the investigator in documenting a fire scene.

FIRE SCENE DOCUMENTATION AND ANALYSIS
Scene Documentation Forms

Types of building construction. Types of construction are shown narratively.

FIRE RESISTIVE CONST. (TYPE I)	WOOD-FRAME CONST. (TYPE V)

Height. Indicate number of stories above ground, number of stories below ground, and height from grade to eaves.

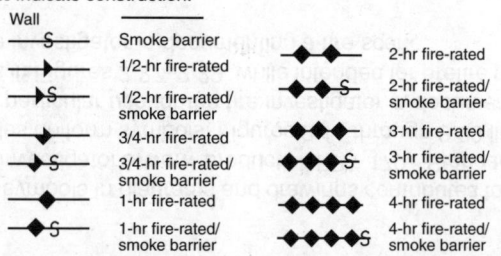

A   Three stories, no basement, 40 ft to eaves.

B   One story with basement, 20 ft to eaves.

C   One equals two stories, no basement, 24 ft to eaves.

D   One-story open porch or shed.

E   One-story addition.

F   Thirteen stories with basement.

G   Underground structure.

(Includes copyrighted material of Insurance Services Office with its permission. Copyright, Insurance Services Office 1975.)

Walls. Indicate construction.

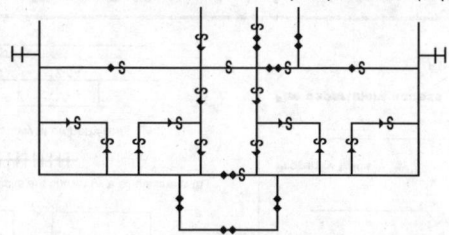

Parapets. The symbol for parapets utilizes one cross for each 6 in. (152 mm) that the parapet extends above the roof. The cross is drawn through an extension of the wall line for the parapeted wall (in plan view).

Symbol used to note wall ratings and parapets on life safety plans and risk analysis plans/cross sections.

Figure 2.11 Symbols for building construction: types of building construction, height, walls, and parapets.

Source: *Fire Protection Handbook,* 19th edition, Figure 3.15.2.

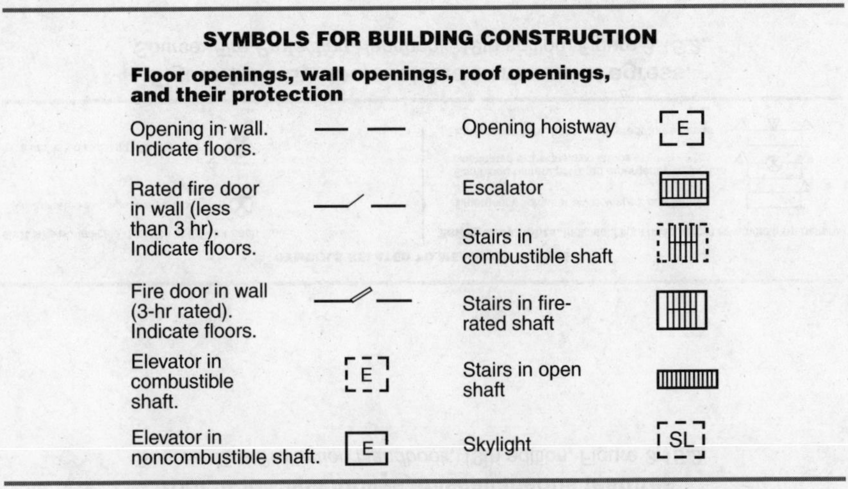

SYMBOLS FOR BUILDING CONSTRUCTION

Floor openings, wall openings, roof openings, and their protection

Opening in wall. Indicate floors.	— —	Opening hoistway	
Rated fire door in wall (less than 3 hr). Indicate floors.		Escalator	
		Stairs in combustible shaft	
Fire door in wall (3-hr rated). Indicate floors.		Stairs in fire-rated shaft	
Elevator in combustible shaft.		Stairs in open shaft	
Elevator in noncombustible shaft.		Skylight	

Figure 2.12 Symbols for building construction: floor openings, wall openings, roof openings, and their protection.

Source: *Fire Protection Handbook,* 19th edition, Figure 3.15.2.

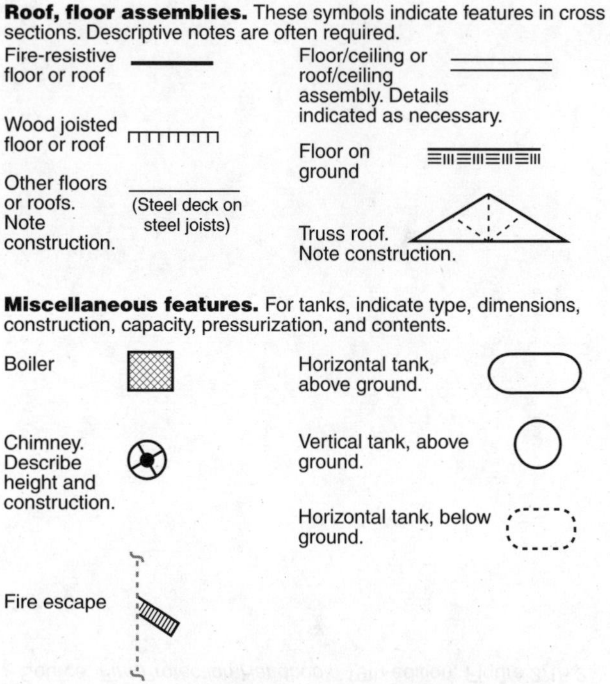

Roof, floor assemblies. These symbols indicate features in cross sections. Descriptive notes are often required.

Fire-resistive floor or roof ⎯⎯⎯

Floor/ceiling or roof/ceiling assembly. Details indicated as necessary. ═══════

Wood joisted floor or roof ⊓⊓⊓⊓⊓⊓⊓

Other floors or roofs. Note construction. ⎯⎯⎯ (Steel deck on steel joists)

Floor on ground ≡⦀≡⦀≡⦀≡⦀

Truss roof. Note construction.

Miscellaneous features. For tanks, indicate type, dimensions, construction, capacity, pressurization, and contents.

Boiler

Horizontal tank, above ground.

Chimney. Describe height and construction.

Vertical tank, above ground.

Horizontal tank, below ground.

Fire escape

Figure 2.13 Symbols for building construction: roof, floor assemblies, miscellaneous features.

Source: *Fire Protection Handbook,* 19th edition, Figure 3.15.2.

SYMBOLS RELATED TO MEANS OF EGRESS

Exit signs. Indicate direction of flow for each face.

Illuminated exit sign, single face

Illuminated exit sign, double face

Emergency lights. Indicate if light head (lamp) is remote from battery.

Emergency light, battery powered, one lamp

Combined battery powered emergency lights and illuminated exit signs, two lamps

Emergency light, battery powered, three lamps

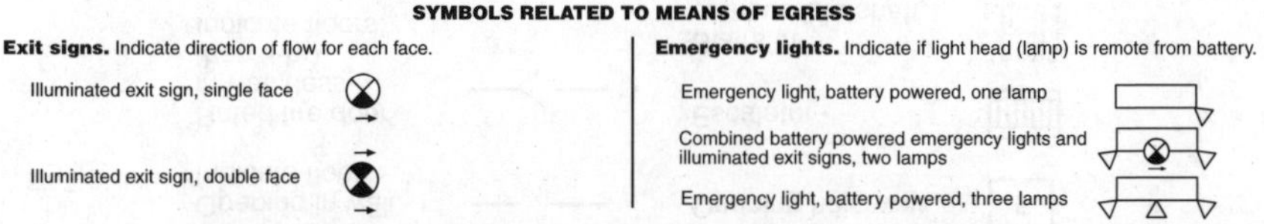

Figure 2.14 Symbols related to means of egress.

Source: *Fire Protection Handbook,* 19th edition, Figure 3.15.2.

SYMBOLS FOR WATER SUPPLY AND DISTRIBUTION
Mains, pipe. Indicate pipe size and material.

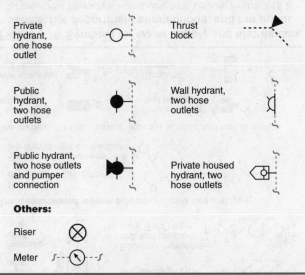

Public water main

Water main under building

Private water main

Suction pipe

Hydrants. Indicate size, type of thread, or connection. Symbol element may be utilized in any combination to fit the type of hydrant.

Private hydrant, one hose outlet

Thrust block

Public hydrant, two hose outlets

Wall hydrant, two hose outlets

Public hydrant, two hose outlets and pumper connection

Private housed hydrant, two hose outlets

Others:

Riser

Meter

Figure 2.15 Symbols for water supply and distribution: mains, pipe, hydrants, and others.
Source: *Fire Protection Handbook,* 19th edition, Figure 3.15.2.

FIRE SCENE DOCUMENTATION AND ANALYSIS
Scene Documentation Forms

SYMBOLS FOR WATER SUPPLY AND DISTRIBUTION

Valves. Indicate valve size.

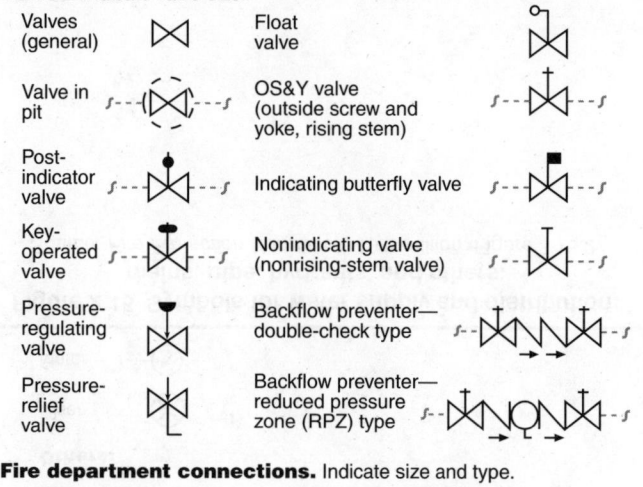

Valves (general)

Valve in pit

Post-indicator valve

Key-operated valve

Pressure-regulating valve

Pressure-relief valve

Float valve

OS&Y valve (outside screw and yoke, rising stem)

Indicating butterfly valve

Nonindicating valve (nonrising-stem valve)

Backflow preventer—double-check type

Backflow preventer—reduced pressure zone (RPZ) type

Fire department connections. Indicate size and type.

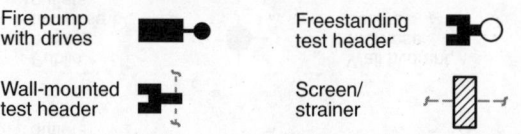

Siamese fire department connection

Free-standing siamese fire department connection

Single fire department connection

Fire pump. For test headers. Indicate number and size of outlets.

Fire pump with drives

Wall-mounted test header

Freestanding test header

Screen/strainer

Figure 2.16 Symbols for water supply and distribution: valves, fire department connections, and fire pump.

Source: *Fire Protection Handbook*, 19th edition, Figure 3.15.2.

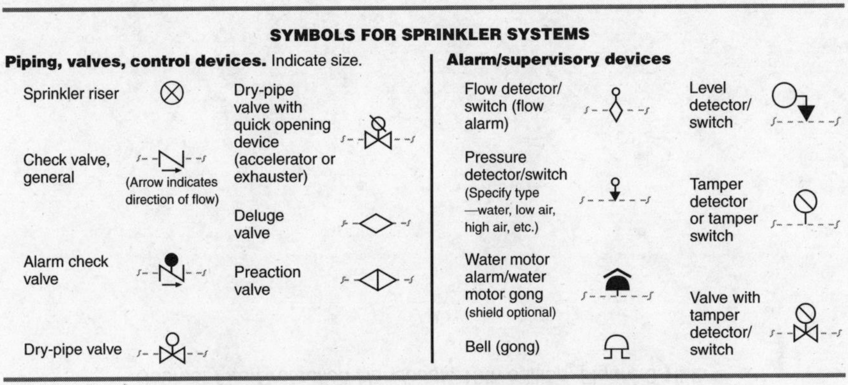

SYMBOLS FOR SPRINKLER SYSTEMS

Piping, valves, control devices. Indicate size.

Sprinkler riser	⊗	Dry-pipe valve with quick opening device (accelerator or exhauster)	
Check valve, general	(Arrow indicates direction of flow)		
		Deluge valve	
Alarm check valve		Preaction valve	
Dry-pipe valve			

Alarm/supervisory devices

Flow detector/ switch (flow alarm)		Level detector/ switch	
Pressure detector/switch (Specify type —water, low air, high air, etc.)		Tamper detector or tamper switch	
Water motor alarm/water motor gong (shield optional)		Valve with tamper detector/ switch	
Bell (gong)			

Figure 2.17 Symbols for sprinkler systems.
Source: *Fire Protection Handbook*, 19th edition, Figure 3.15.2.

SYMBOLS FOR FIRE SPRINKLERS

Upright sprinkler

Note "DP" on drawing and/or in specifications where dry pendent sprinklers are employed.

Upright sprinkler, nippled up

Sprinkler, with guard

Upright sprinkler head shown

Outside sprinkler

Specify type, orifice size, for example: Open sprinkler (window or cornice).

Pendent sprinkler, on drop nipple

Note "DP" on drawing and/or in specifications where dry pendent sprinklers are employed.

Pendent sprinkler

Sidewall sprinkler

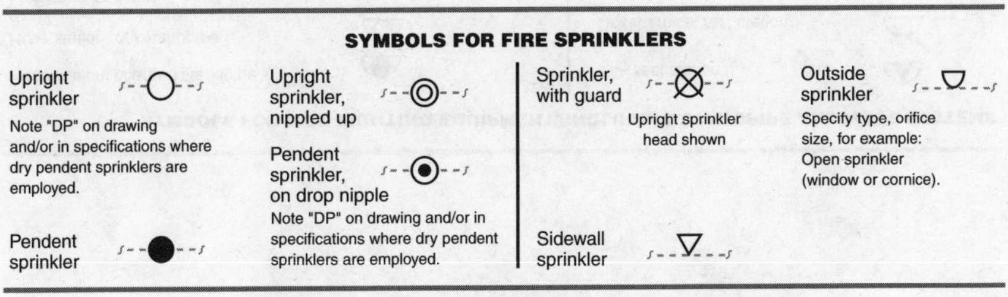

Figure 2.18 Symbols for fire sprinklers.
Source: *Fire Protection Handbook*, 19th edition, Figure 3.15.2.

SYMBOLS FOR PIPING, VALVES, CONTROL DEVICES, AND HANGERS

Sprinkler piping and branch line

Indicate pipe size.

Pipe hanger

This symbol is a diagonal stroke imposed on the pipe that it supports.

Angle valve (angle hose valve)

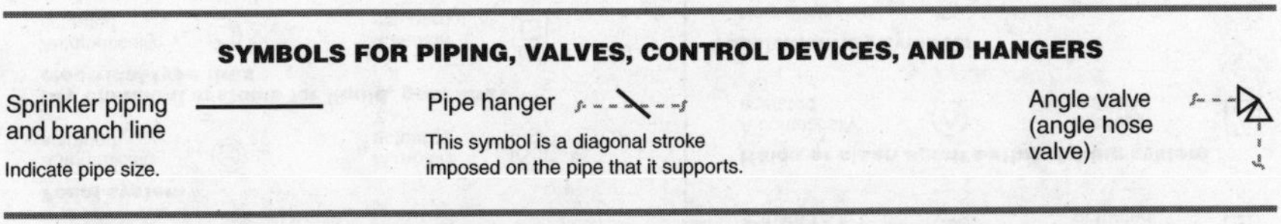

Figure 2.19 Symbols for piping, valves, control devices, and hangers.
Source: *Fire Protection Handbook*, 19th edition, Figure 3.15.2.

FIRE SCENE DOCUMENTATION AND ANALYSIS
Scene Documentation Forms

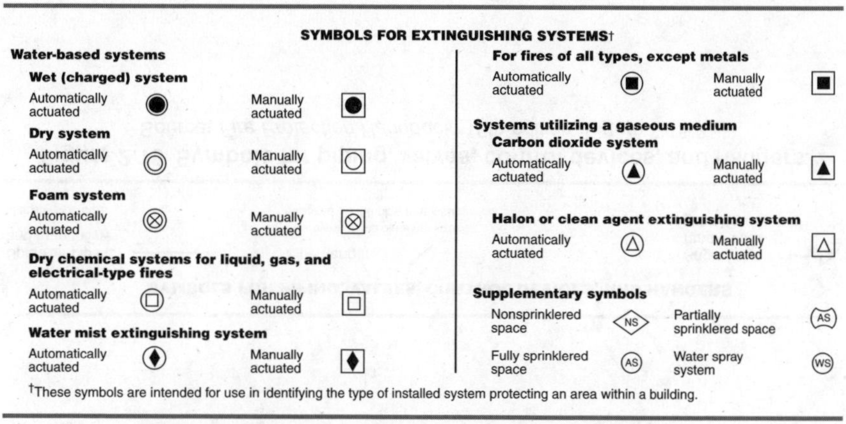

FIRE SCENE DOCUMENTATION AND ANALYSIS
Scene Documentation Forms

SYMBOLS FOR EXTINGUISHING SYSTEMS†

Water-based systems

Wet (charged) system

| Automatically actuated | ● | Manually actuated | ◉ |

Dry system

| Automatically actuated | ○ | Manually actuated | ◻○ |

Foam system

| Automatically actuated | ⊗ | Manually actuated | ⊠ |

Dry chemical systems for liquid, gas, and electrical-type fires

| Automatically actuated | ⊡ | Manually actuated | ▢ |

Water mist extinguishing system

| Automatically actuated | ◆ | Manually actuated | ◈ |

For fires of all types, except metals

| Automatically actuated | ▣ | Manually actuated | ▪ |

Systems utilizing a gaseous medium
Carbon dioxide system

| Automatically actuated | ▲ | Manually actuated | ▲ |

Halon or clean agent extinguishing system

| Automatically actuated | △ | Manually actuated | △ |

Supplementary symbols

| Nonsprinklered space | NS | Partially sprinklered space | AS |
| Fully sprinklered space | AS | Water spray system | WS |

†These symbols are intended for use in identifying the type of installed system protecting an area within a building.

Figure 2.20 Symbols for extinguishing systems.
Source: *Fire Protection Handbook*, 19th edition, Figure 3.15.2.

SYMBOLS FOR FIRE-FIGHTING EQUIPMENT, INCLUDING STANDPIPE AND HOSE SYSTEMS

Hose station, charged standpipe		CO₂ reel station	
Hose station, dry standpipe		Dry chemical reel station	
Monitor nozzle, dry		Foam reel station	
Monitor nozzle, charged			

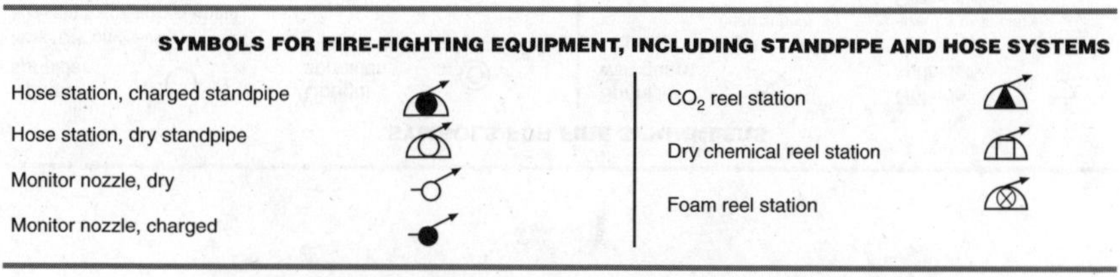

Figure 2.21 Symbols for fire-fighting equipment, including standpipe and hose systems.
Source: *Fire Protection Handbook*, 19th edition, Figure 3.15.2.

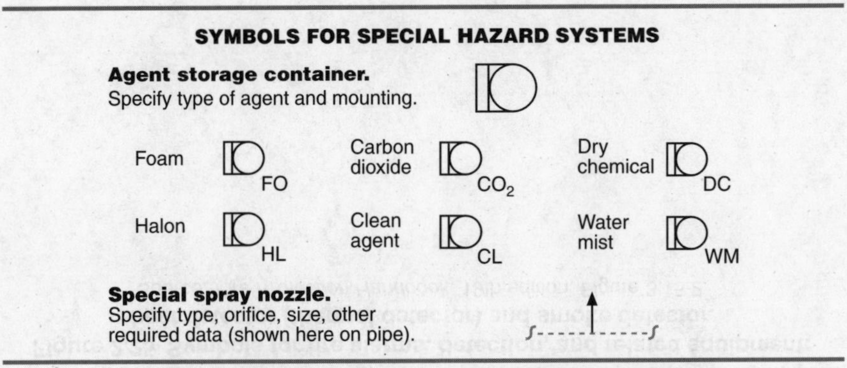

SYMBOLS FOR SPECIAL HAZARD SYSTEMS

Agent storage container.
Specify type of agent and mounting.

Foam · FO

Carbon dioxide · CO_2

Dry chemical · DC

Halon · HL

Clean agent · CL

Water mist · WM

Special spray nozzle.
Specify type, orifice, size, other
required data (shown here on pipe).

Figure 2.22 Symbols for special hazard systems.
Source: *Fire Protection Handbook*, 19th edition, Figure 3.15.2.

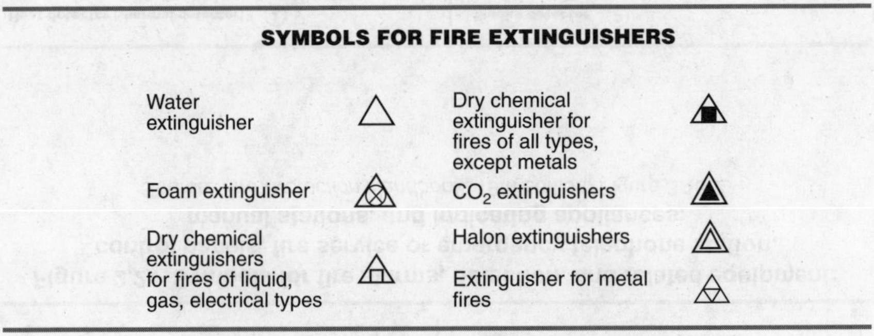

SYMBOLS FOR FIRE EXTINGUISHERS

Water extinguisher

Dry chemical extinguisher for fires of all types, except metals

Foam extinguisher

CO_2 extinguishers

Dry chemical extinguishers for fires of liquid, gas, electrical types

Halon extinguishers

Extinguisher for metal fires

Figure 2.23 Symbols for fire extinguishers.
Source: *Fire Protection Handbook,* Figure 3.15.2.

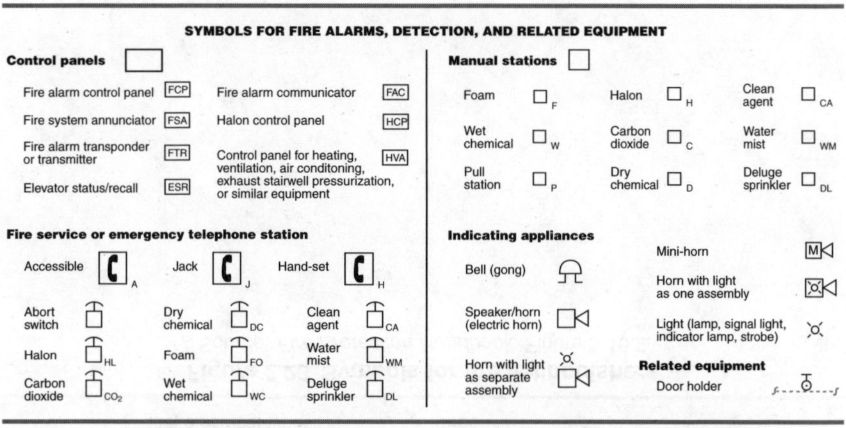

SYMBOLS FOR FIRE ALARMS, DETECTION, AND RELATED EQUIPMENT

Control panels ☐

Fire alarm control panel FCP

Fire system annunciator FSA

Fire alarm transponder or transmitter FTR

Elevator status/recall ESR

Fire alarm communicator FAC

Halon control panel HCP

Control panel for heating, ventilation, air conditioning, exhaust stairwell pressurization, or similar equipment HVA

Manual stations ☐

Foam ☐F

Wet chemical ☐W

Pull station ☐P

Halon ☐H

Carbon dioxide ☐C

Dry chemical ☐D

Clean agent ☐CA

Water mist ☐WM

Deluge sprinkler ☐DL

Fire service or emergency telephone station

Accessible **C**A

Abort switch ⌂

Halon ⌂HL

Carbon dioxide ⌂CO₂

Jack **C**J

Dry chemical ⌂DC

Foam ⌂FO

Wet chemical ⌂WC

Hand-set **C**H

Clean agent ⌂CA

Water mist ⌂WM

Deluge sprinkler ⌂DL

Indicating appliances

Bell (gong) ⌂

Speaker/horn (electric horn) ◁

Horn with light as separate assembly ◁

Mini-horn M◁

Horn with light as one assembly ◁

Light (lamp, signal light, indicator lamp, strobe) ☼

Related equipment

Door holder ⌂

Figure 2.24 Symbols for fire alarms, detection, and related equipment: control panels, fire service or emergency telephone station, manual stations, and indicating appliances.
Source: *Fire Protection Handbook*, 19th edition, Figure 3.15.2.

Heat detector (thermal detector) ⌂

Combination — rate of rise and fixed temperature ⌂R/F

Rate compensation ⌂R/C

Fixed temperature ⌂F

Line-type detector (heat-sensitive cable) ⌂→

Rate of rise only ⌂R

Gas detector ▲

Flame detector (flicker detector) ⟨∧⟩

Indicate UV, IR, or visible radiation type ⟨∧⟩

Smoke detector ⌀

Photoelectric products of combustion detector ⌀P

Ionization products of combustion detector ⌀I

Beam transmitter ⌀BT

Beam receiver ⌀BR

Smoke detector in duct ⌀

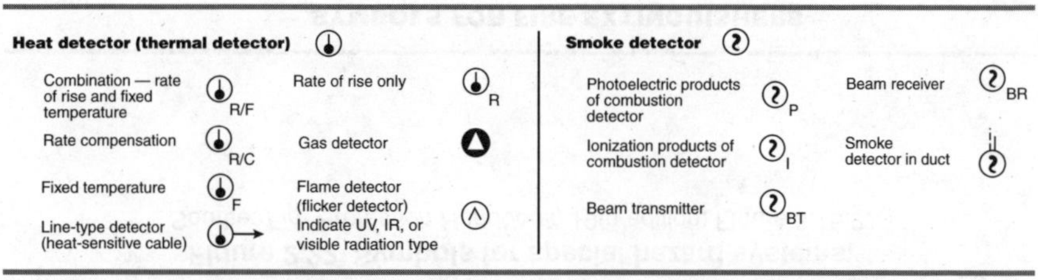

Figure 2.25 Symbols for fire alarms, detection, and related equipment: heat detector (thermal detector) and smoke detector.
Source: *Fire Protection Handbook*, 19th edition, Figure 3.15.2.

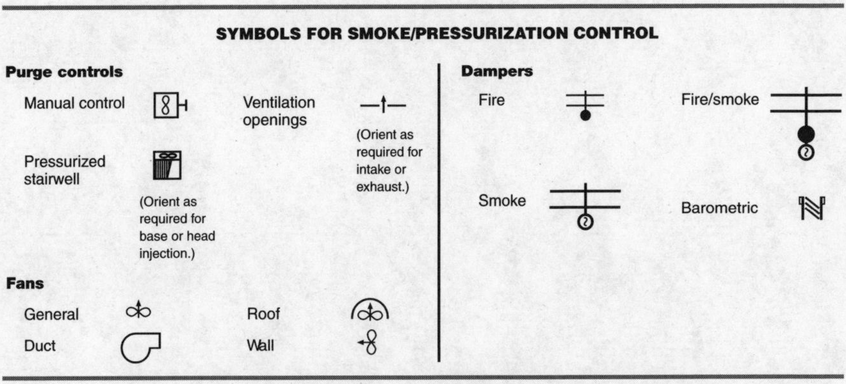

SYMBOLS FOR SMOKE/PRESSURIZATION CONTROL

Purge controls

Manual control

Ventilation openings

(Orient as required for intake or exhaust.)

Pressurized stairwell

(Orient as required for base or head injection.)

Dampers

Fire

Fire/smoke

Smoke

Barometric

Fans

General

Roof

Duct

Wall

Figure 2.26 Symbols for smoke/pressurization control.
Source: *Fire Protection Handbook*, 19th edition, Figure 3.15.2.

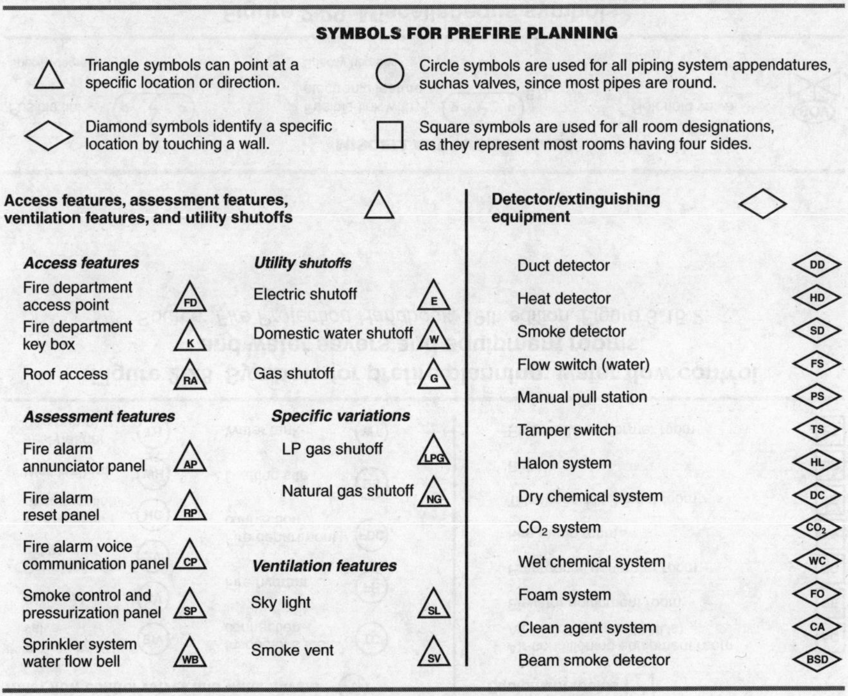

SYMBOLS FOR PREFIRE PLANNING

Triangle symbols can point at a specific location or direction.

Circle symbols are used for all piping system appendatures, suchas valves, since most pipes are round.

Diamond symbols identify a specific location by touching a wall.

Square symbols are used for all room designations, as they represent most rooms having four sides.

Access features, assessment features, ventilation features, and utility shutoffs

Detector/extinguishing equipment

Access features

Fire department access point — FD

Fire department key box — K

Roof access — RA

Assessment features

Fire alarm annunciator panel — AP

Fire alarm reset panel — RP

Fire alarm voice communication panel — CP

Smoke control and pressurization panel — SP

Sprinkler system water flow bell — WB

Utility shutoffs

Electric shutoff — E

Domestic water shutoff — W

Gas shutoff — G

Specific variations

LP gas shutoff — LPG

Natural gas shutoff — NG

Ventilation features

Sky light — SL

Smoke vent — SV

Duct detector — DD

Heat detector — HD

Smoke detector — SD

Flow switch (water) — FS

Manual pull station — PS

Tamper switch — TS

Halon system — HL

Dry chemical system — DC

CO_2 system — CO_2

Wet chemical system — WC

Foam system — FO

Clean agent system — CA

Beam smoke detector — BSD

Figure 2.27 Symbols for prefire planning.
Source: *Fire Protection Handbook*, Figure 3.15.2.

FIRE SCENE DOCUMENTATION AND ANALYSIS
Scene Documentation Forms

Water flow control valves and water savers ◯

Post-indicator valve	(PIV)	Inspector's test connection	(TC)
Riser valve	(RV)	Fire hydrant	(FH)
Sprinkler zone valve	(ZV)		
Hose cabinet or connection	(HC)	Fire department connection	(FDC)
Wall hydrant	(WH)	Drafting site	(DS)
Test header (fire pump)	(TH)	Water tank	(WT)

Equipment rooms ▢

Air-conditioning equipment room Air-handling units (AHUs)	AC
Elevator equipment room	EE
Emergency generator room	EG
Fire pump room	FP
Telephone equipment room	TE
Boiler room	BR
Electrical/transformer room	ET

Figure 2.28 Symbols for prefire planning: water flow control and water savers and equipment rooms.

Source: *Fire Protection Handbook*, 19th edition, Figure 3.15.2.

MISCELLANEOUS SYMBOLS

Fusible link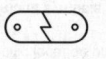

Specify degrees.

Fusible link with electronic feature ⊙⧖⊙ ETL

Specify degrees.

Solenoid valve (SOV)⧓

Figure 2.29 Miscellaneous symbols.

Source: *Fire Protection Handbook*, 19th edition, Figure 3.15.2.

Table 2.2 Color Code for Denoting Construction Materials for Walls

Color	Interpretation
Brown	Fire-resistive protected steel
Red	Brick, hollow tile
Yellow	Frame-wood, stucco
Blue	Concrete, stone, or hollow concrete block
Gray	Noncombustible (sheet metal or metal lath and plaster) unprotected steel

Source: *Fire Protection Handbook,* 19th edition, Table 3.15.2.

For fire investigators who have access to color printers, specific colors can be used to identify construction materials in walls. Table 2.2 presents those color codes.

In structural fire investigation, the investigator will often prepare floor plans of the room or building. Figure 2.30 illustrates the *minimum* drawing needed for simple fire analysis.

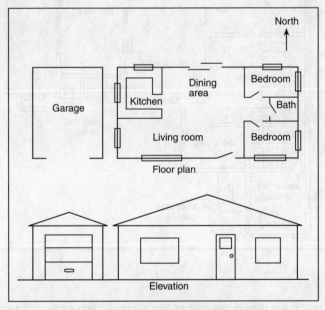

Figure 2.30 Minimum drawing for simple fire analysis.
Source: NFPA 921, 2001 edition, Figure 13.4.5.

In addition to minimum drawings, the fire investigator will sketch the contents of the room or building and, in the case of fatal fires, diagram the location of fire victims. Figures 2.31–2.39 can assist the investigator with these fire scene tasks.

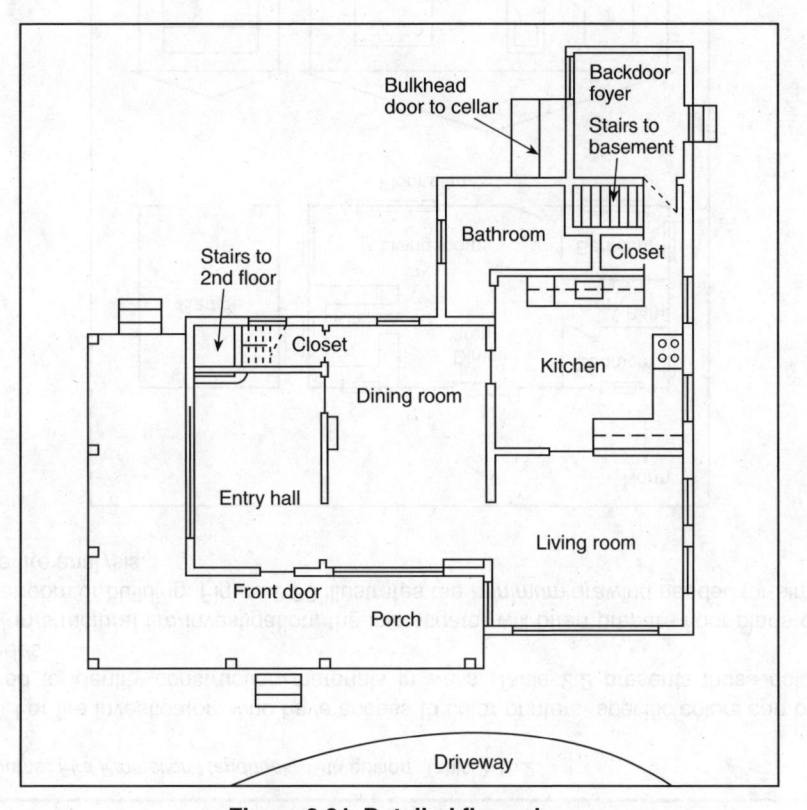

Figure 2.31 Detailed floor plan.
Source: NFPA 921, 2001 edition, Figure 13.4.2(b).

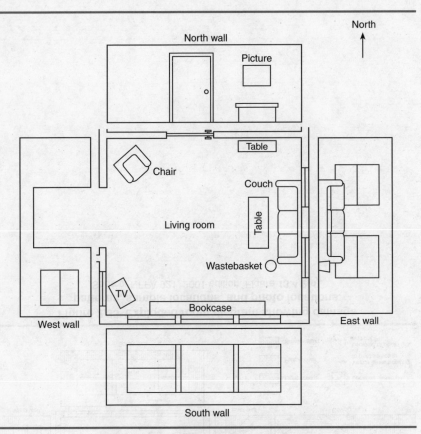

Figure 2.32 Pre-fire contents diagram.
Source: NFPA 921, 2001 edition, Figure 13.4.2(c).

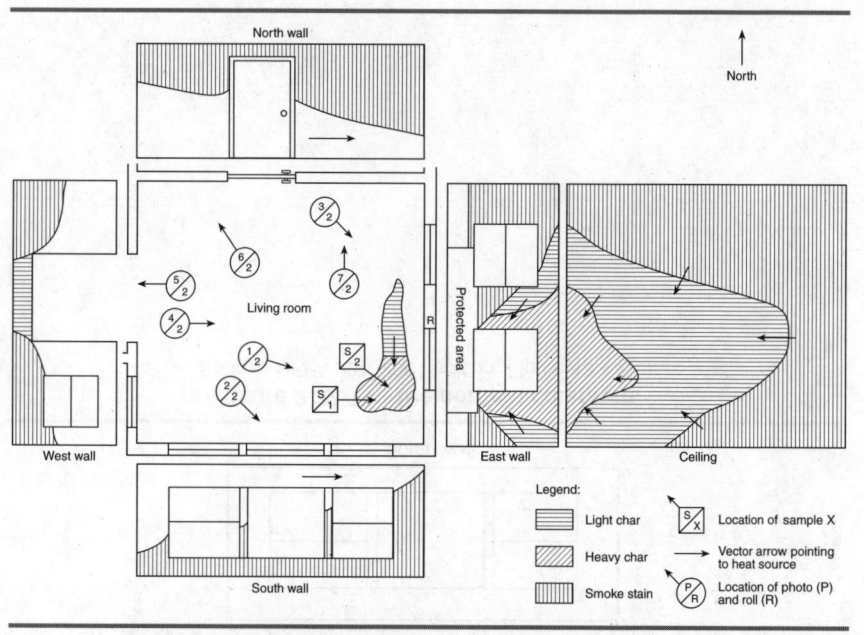

Figure 2.33 Exploded room diagram showing damage patterns, sample locations, and photo locations.
Source: NFPA 921, 2001 edition, Figure 13.4.2(d).

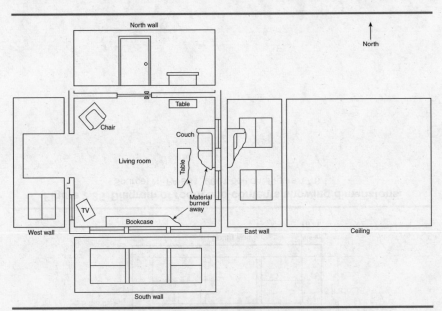

**Figure 2.34 Contents reconstruction diagram showing
damaged furniture in original positions.**
Source: NFPA 921, 2001 edition, Figure 13.4.2(e).

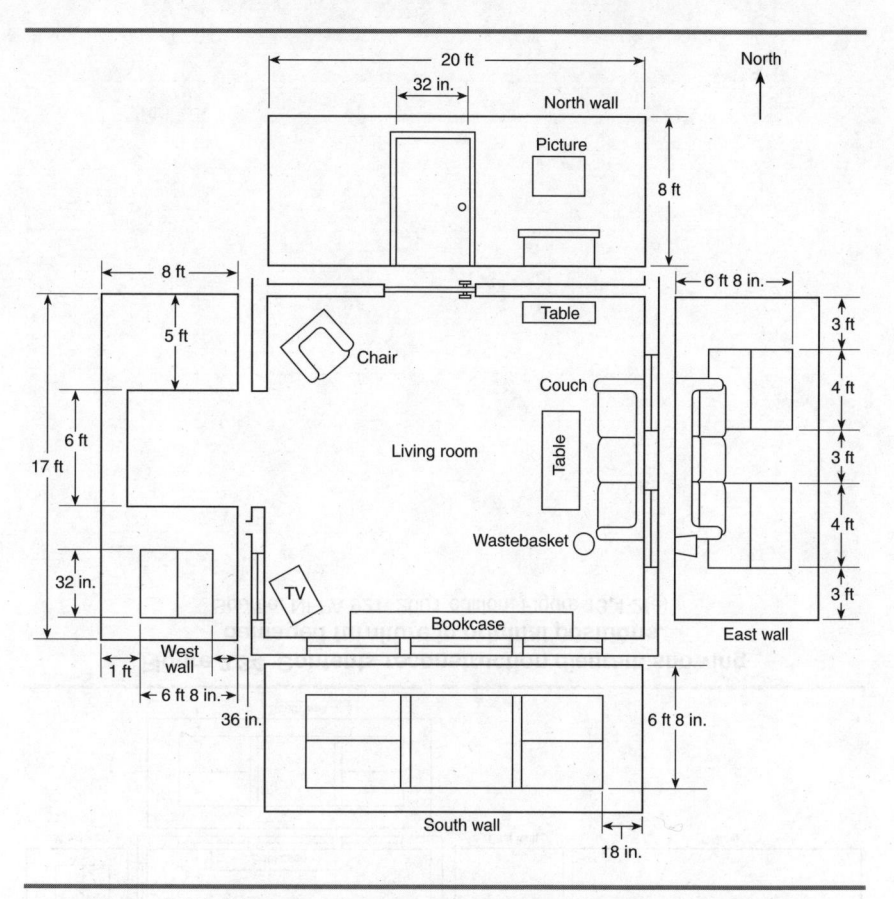

Figure 2.35 Diagram of room and contents showing dimensions.
Source: NFPA 921, 2001 edition, Figure 17.6.1.

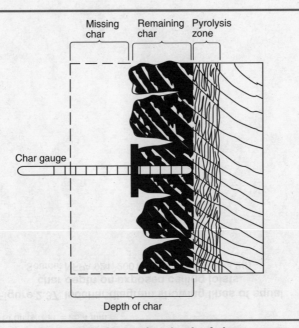

Figure 2.36 Measuring depth of char.
Source: NFPA 921, 2001 edition, Figure 4.5.3.3(a).

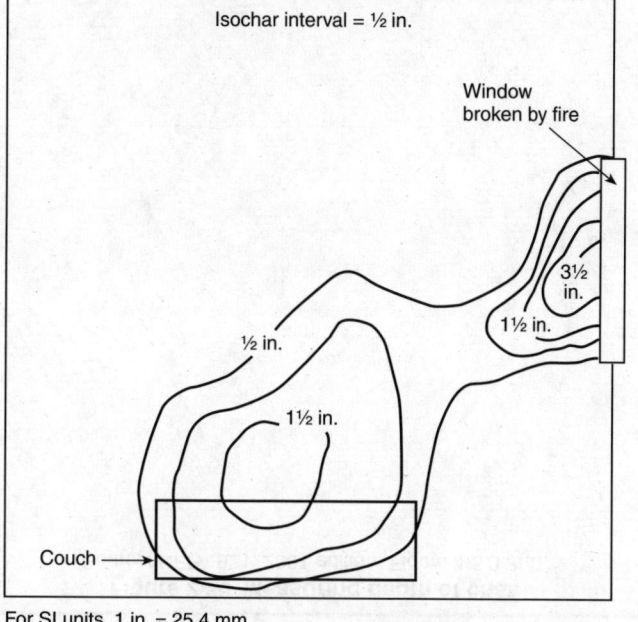

For SI units, 1 in. = 25.4 mm.

Figure 2.37 Isochar diagram showing lines of equal char depth on exposed ceiling joists.
Source: NFPA 921, 2001 edition, Figure 13.4.2(f).

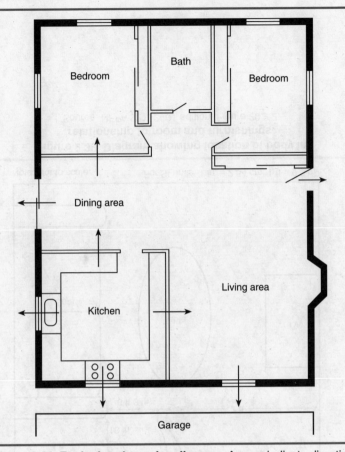

Figure 2.38 Explosion dynamics diagram. Arrows indicate direction
of displacement of walls, doors, and windows.
Source: NFPA 921, 2001 edition, Figure 18.14.

FIRE SCENE DOCUMENTATION AND ANALYSIS
Scene Documentation Forms

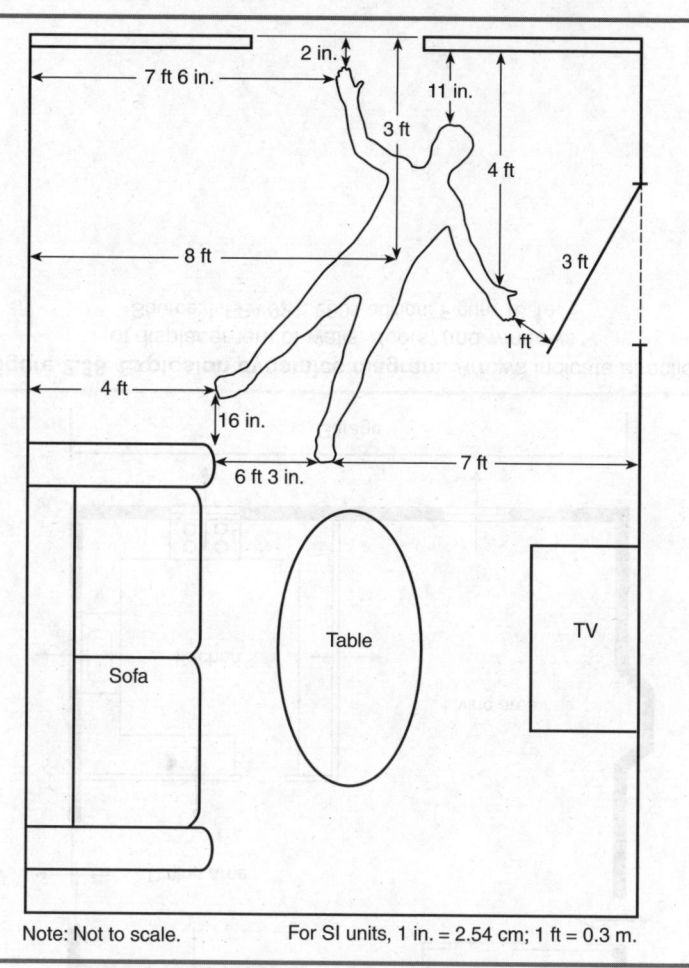

Note: Not to scale.   For SI units, 1 in. = 2.54 cm; 1 ft = 0.3 m.

Figure 2.39 Diagram showing location of body in relationship to room and furnishings.
Source: NFPA 921, 2001 edition, Figure 20.2.2.

SAMPLE DOCUMENTATION PHOTOGRAPHS

Visual impressions of the fire scene can best be documented using photographs and video recording. Although for still photography most investigators use 35-millimeter color film and single lens reflex cameras, simple viewfinder cameras or even single use cameras are better than no photos at all. In fact, many investigators will carry a simple camera as a back up in the event that the "big rig" fails or the batteries go dead. Individual investigators often have their own systematic approach to photographing or videotaping the fire scene.

Digital photography is now becoming more and more common. While some investigators have switched completely, others use film as a back up.

Regardless of the type of still or video camera used, what is important to remember is that the best documentation will provide overlapping shots so that there will be continuous coverage. Often neglected areas are ceilings and areas behind doors. When taking close up shots of evidence or patterns there should also be an overview shot taken that shows the location of the close up relative to the areas around it. It is difficult to suggest just how much visual documentation is enough. Evidence that leads to identification of the area and point of origin should be fully documented. Evidence supporting a proposed cause, and evidence ruling out other possible causes, should also be fully documented to the extent possible.

Figures 2.40 through 2.44 provide examples of photo documentation forms and suggested shooting angles.

Photo Sequence

When documenting the scene with photographs, it is important to have as much coverage as possible, and in particular overlapping photos and a photo log and photo diagram should be prepared. It is often helpful to have a routine for the initial documentation process such as beginning with a long distance shot or two showing the building(s) in context with the neighborhood, followed by shots around the outside building showing the sides and corners. Overlapping inside shots should cover the walls, floors, ceilings, and both sides of doors.

One technique to document a large scene or an interior area that won't fit on a single photo and still show the needed detail is to take overlapping shots with "key elements" that appear in the overlap areas of adjacent photos. The shots can then be matched up manually for film media or "stitched" together electronically for digital media to form a continuous picture called a collage or panorama. When there is a small area of interest and a close up shot is desired, several shots can be taken zooming in to the subject of interest so that the location of the subject can be seen in context of the larger area in the longer shots. Figures 2.45 and 2.46 illustrate collage/panorama shots and zoom sequences.

FIRE SCENE DOCUMENTATION AND ANALYSIS
Sample Documentation Photographs

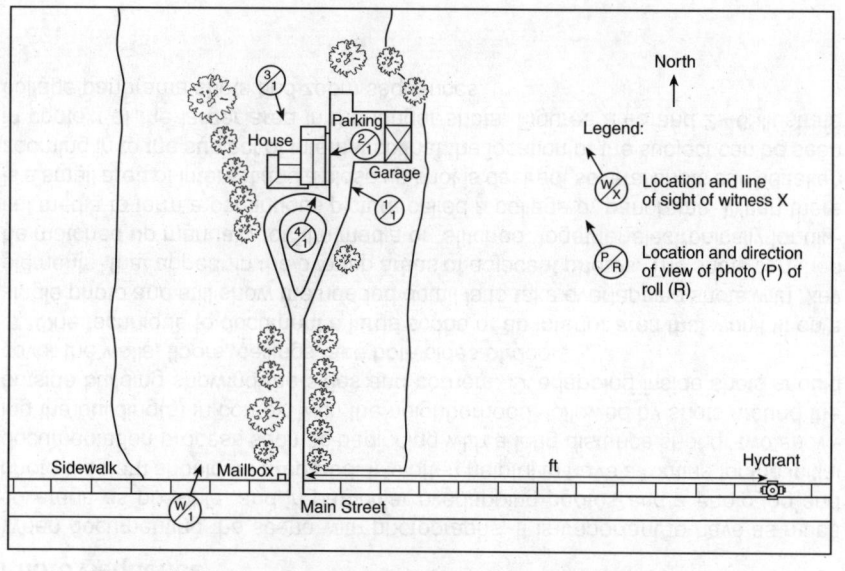

Figure 2.40 Site plan showing photo and witness locations.
Source: NFPA 921, Figure 13.4.2(a).

PHOTOGRAPH LOG

Roll #   Exposures:

Case #	Date:		
Camera Make/Type:	Film Type:	Film Speed:	ASA

Number	Description	Location
1)		
2)		
3)		
4)		
5)		
6)		
7)		
8)		
9)		
10)		
11)		
12)		
13)		
14)		
15)		
16)		
17)		
18)		
19)		
20)		
21)		
22)		
23)		
24)		
25)		
26)		
27)		
28)		
29)		
30)		
31)		
32)		
33)		
34)		
35)		
36)		

Photos taken by: _____   Initials: _____

Figure 2.41 Sample photograph log.

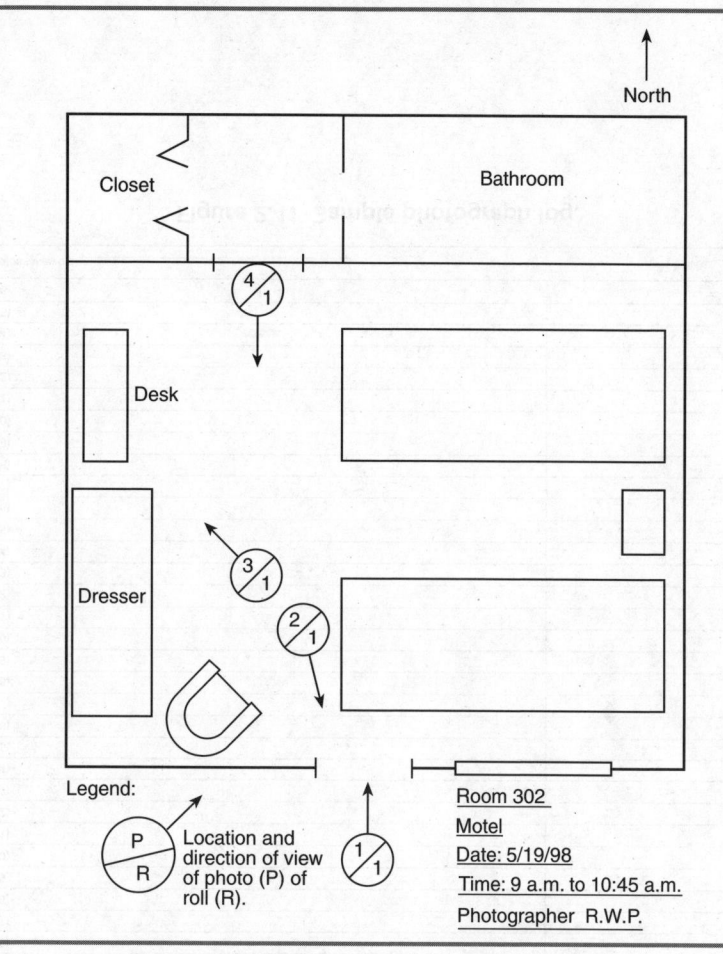

Figure 2.42 Diagram showing photo locations.
Source: NFPA 921, 2001 edition, Figure 13.2.3.3.

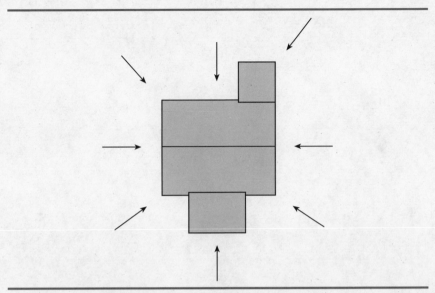

Figure 2.43 Photographing the scene from all angles and corners.
Source: NFPA 921, 2001 edition, Figure 13.2.5.4.

FIRE SCENE DOCUMENTATION AND ANALYSIS
Sample Documentation Photographs

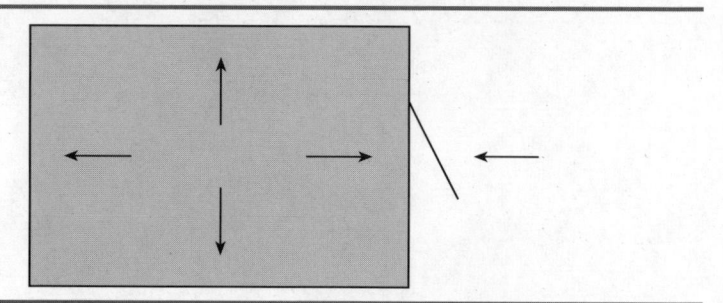

Figure 2.44 Photographing all four walls and both sides of each door.
Source: NFPA 921, 2001 edition, Figure 13.2.5.6(a).

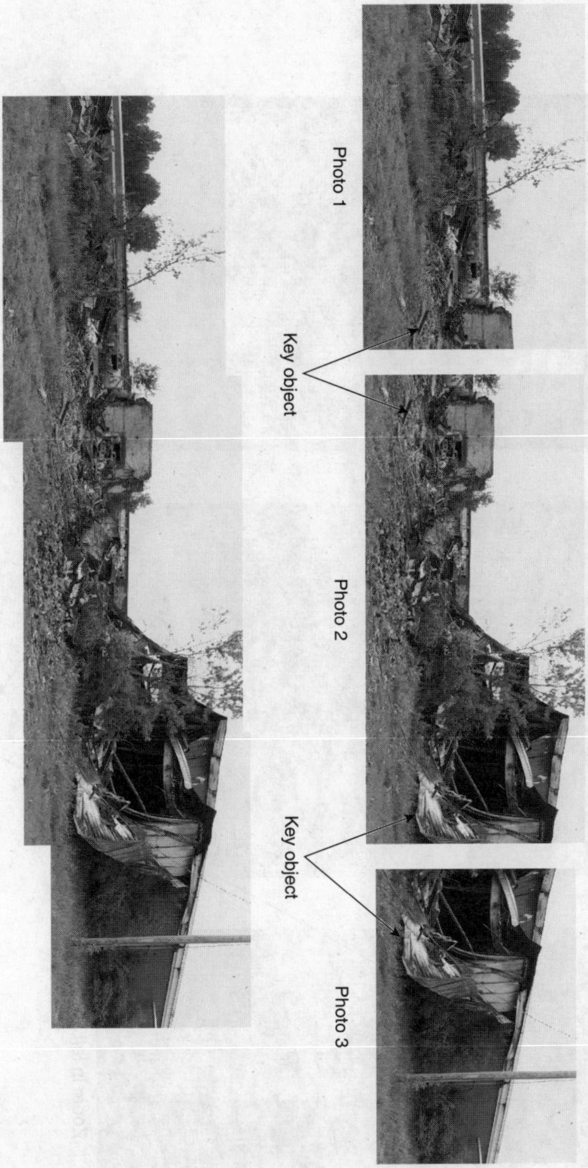

Figure 2.45 Example of a panorama or collage, in sections and assembled.

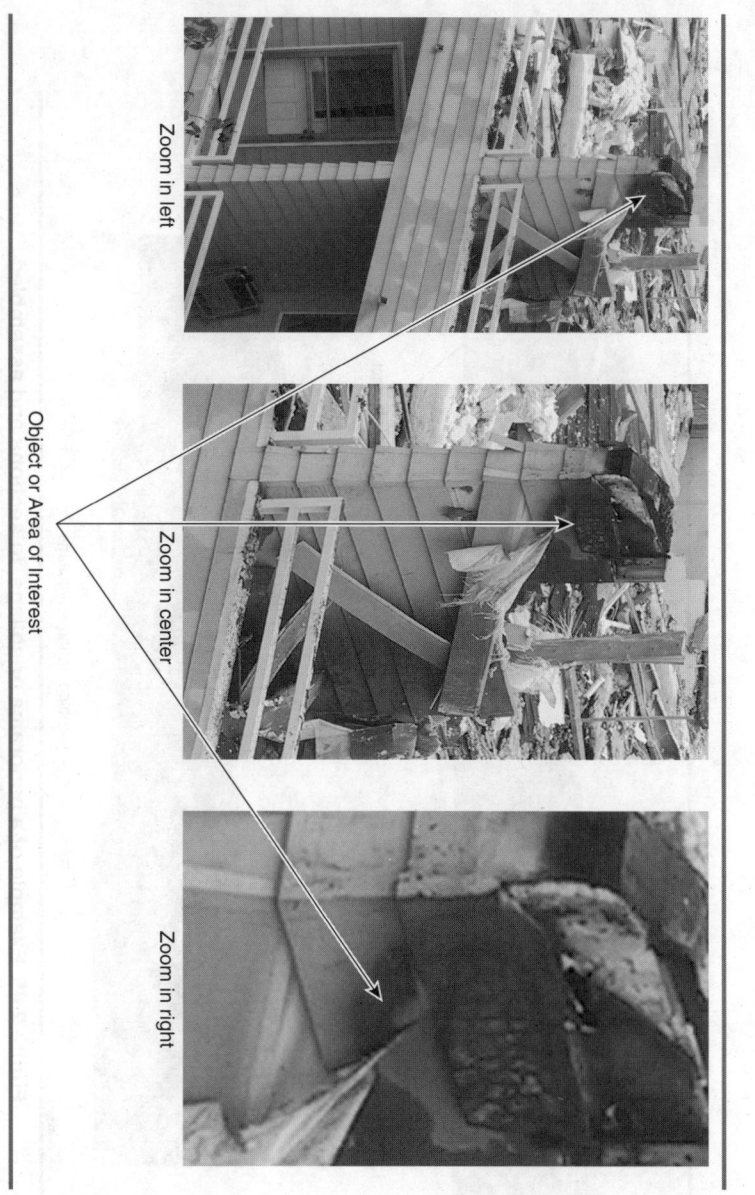

Figure 2.46 Example of "zoom-in" on object of interest.

BURN PATTERN IDENTIFICATION

Structures and Their Contents

The action of heat and smoke in fires results in patterns on building components such as floors, doors, windows, structural elements, walls, and ceilings. These patterns can take the form of burned or consumed (missing) material, smoke deposits or undamaged areas that were protected during the fire by furnishings or other objects. Patterns can also be present on furnishings and other building contents. Examples of many of these patterns are presented below to assist the investigator in recognizing them in the field. Details regarding the origin and interpretation of these patterns and selected references can be found in NFPA 921.

Figures 2.47 through 2.63 illustrate burn pattern identification in structures and contents.

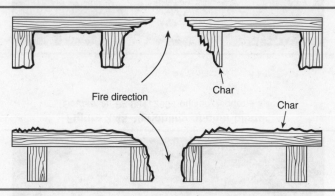

Figure 2.47 Burn pattern with fire from above and below.
Source: NFPA 921, 2001 edition, Figure 4.3.3.

Figure 2.48 Variability of char blister.
Source: NFPA 921, 2001 edition, Figure 4.5.5.

Figure 2.49 Spalling on a ceiling.
Source: NFPA 921, 2001 edition, Figure 4.6.

Figure 2.50 Clean burn on wall surface.
Source: NFPA 921, 2001 edition, Figure 4.11.

Figure 2.51 Typical V pattern showing wall and wood stud damage.
Source: NFPA 921, 2001 edition, Figure 4.17.1.

Figure 2.52 Inverted cone pattern fueled by a natural gas leak below the floor level.
Source: NFPA 921, 2001 edition, Figure 4.17.2.2.

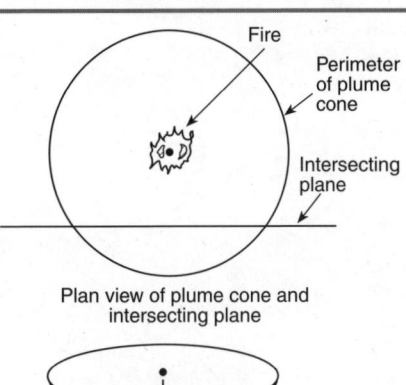

Fire

Perimeter
of plume
cone

Intersecting
plane

Plan view of plume cone and
intersecting plane

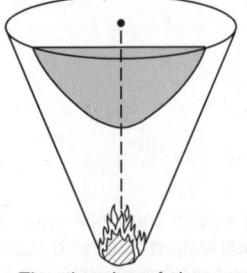

Elevation view of plume cone
showing U-shaped pattern

Figure 2.53 Development of U-shaped pattern.
Source: NFPA 921, 2001 edition, Fig, 4.17.4.

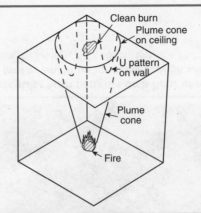

Clean burn
Plume cone
on ceiling
U pattern
on wall
Plume cone
Fire

Figure 2.54 Truncated cone pattern.
Source: NFPA 921, 2001 edition, Figure 4.17.5.

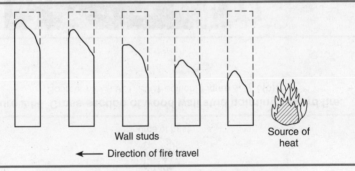

Wall studs

← Direction of fire travel

Source of heat

Figure 2.55 Wood studs showing decreasing damage as distance from fire increases.
Source: NFPA 921, 2001 edition, Figure 4.17.6(a).

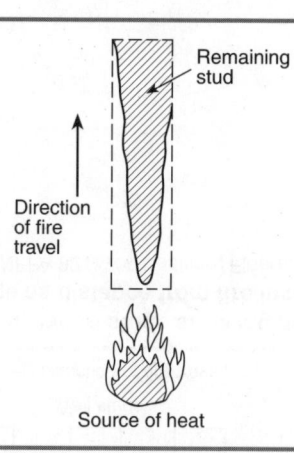

Figure 2.56 Cross section of wood wall stud pointing toward fire.
Source: NFPA 921, 2001 edition, Figure 4.17.6(b).

**Figure 2.57 Irregular burn patterns on a floor of a room burned
in a fire test in which no ignitable liquids were used.**
Source: NFPA 921, 2001 edition, Figure 4.17.7.2(a).

Figure 2.58 Irregularly shaped pattern on floor carpeting resulting from poured ignitable liquid. Burned match can be seen at lower left.
Source: NFPA 921, 2001 edition, Figure 4.17.7.2(b).

Figure 2.59 "Pool-shaped" burn pattern produced by a cardboard box burning on an oak parquet floor.
Source: NFPA 921, 2001 edition, Figure 4.17.7.2(c).

Figure 2.60 Saddle burn in a floor joist.
Source: NFPA 921, 2001 edition, Figure 4.17.10.

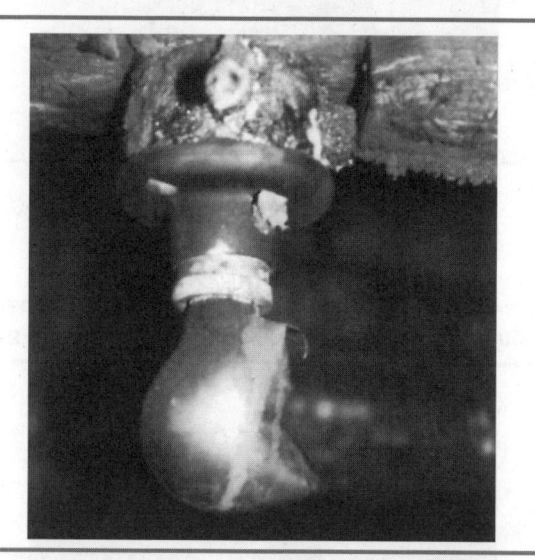

**Figure 2.61 A typical "pulled" bulb showing that
the heating was from the right side.**
Source: NFPA 921, 2001 edition, Figure 4.20.1.

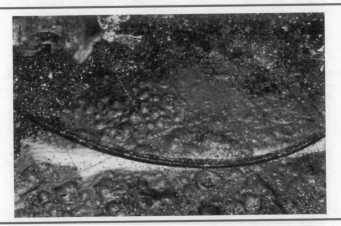

Figure 2.62 Floor tile protected from radiant heat by wire.
Source: NFPA 921, 2001 edition, Figure 13.2.5.6(b).

Figure 2.63 Protected area where body was located.
Source: NFPA 921, 2001 edition, Figure 13.2.5.9.

Vehicles

Many of the patterns associated with building and contents fires are also associated with vehicle fires, particularly motor homes and recreational vehicles. Other patterns are formed by the effects of heat on body panels and windows. Examples of damage patterns associated with vehicle fires are presented below to assist in their recognition. Details regarding origin and interpretation and selected references are presented in NFPA 921.

Figures 2.64 through 2.69 illustrate typical burn patterns in vehicles.

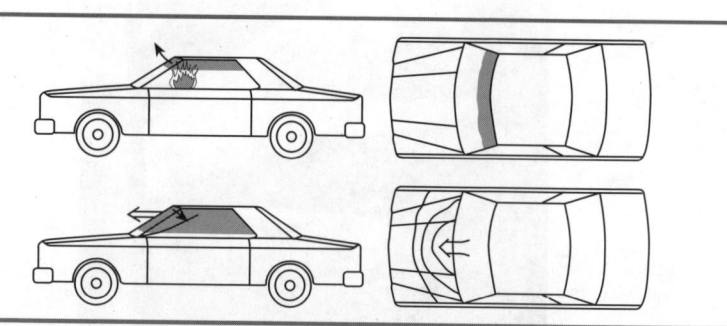

Figure 2.64 Burn pattern development from an interior origin.
Source: NFPA 921, 2001 edition, Figure 22.8(a).

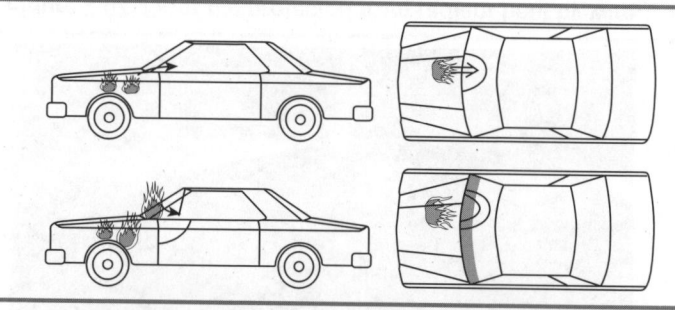

Figure 2.65 Burn pattern development from an engine compartment origin.
Source: NFPA 921, 2001 edition, Figure 22.8(b).

Figure 2.66 Radial burn pattern produced by a passenger compartment fire.
Source: NFPA 921, 2001 edition, Figure 22.8(c).

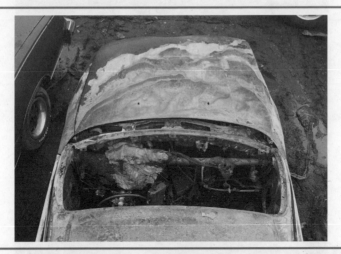

Figure 2.67 Radial burn pattern produced by a passenger compartment fire.
Source: NFPA 921, 2001 edition, Figure 22.8(d).

**Figure 2.68 Incipient windshield failure caused
by an engine compartment fire.**
Source: NFPA 921, 2001 edition, Figure 22.8(e).

**Figure 2.69 Radial pattern on driver's side door produced
by an engine compartment fire.**
Source: NFPA 921, 2001 edition, Figure 22.8(f).

ANALYTICAL TOOLS FOR THE FIRE INVESTIGATOR

Organizing, evaluating, and analyzing the data collected in a fire investigation is an important part of forming the foundation for determination of the fire origin and cause. A number of aids are available to assist in this process. The scientific method is an important part of the process in that possible origin and cause theories that are developed can be evaluated or tested against the collected evidence. This section provides examples of various analytical tools and their applications and an example form for collecting data for use in computer modeling or analysis.

Figures 2.70 through 2.75 and Table 2.3 are sample analytical aids used in fire investigation.

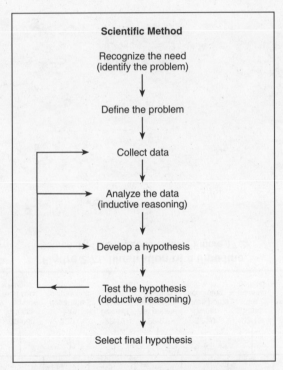

Figure 2.70 Use of the scientific method.
Source: NFPA 921, 2001 edition, Figure 2.3.

FIRE SCENE DOCUMENTATION AND ANALYSIS
Analytical Tools for the Fire Investigator

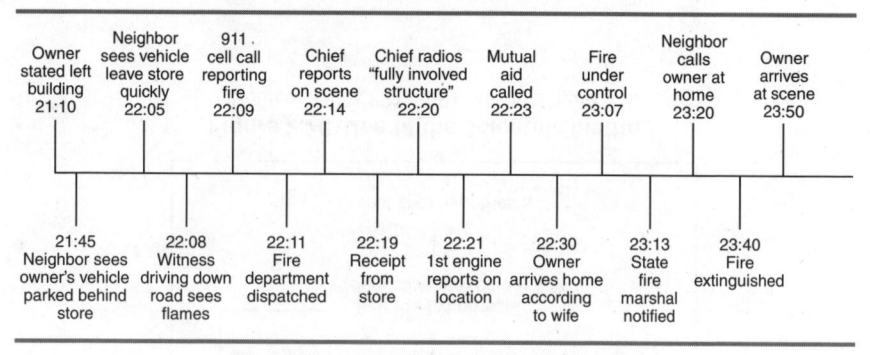

Figure 2.71 Illustration of a time line.
Source: NFPA 921, 2001 edition, Figure 17.2.

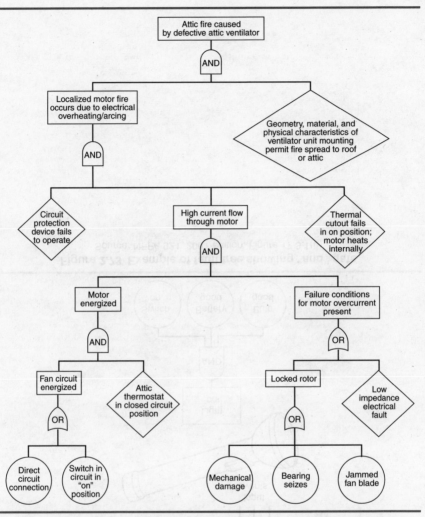

Figure 2.72 Fault tree showing combination of "and" and "or" gates.
Source: NFPA 921, 2001 edition, Figure 17.3.1(a).

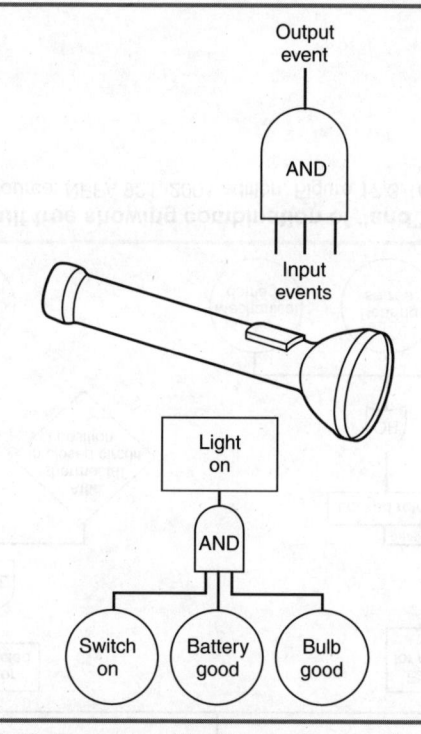

Figure 2.73 Example of fault tree showing "and" gate.
Source: NFPA 921, 2001 edition, Figure 17.3.1(b).

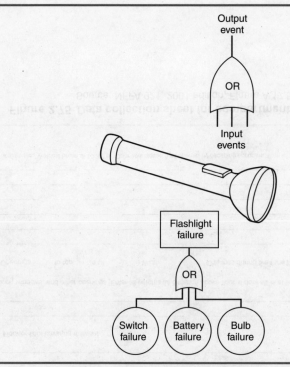

Figure 2.74 Example of fault tree showing "or" gate.
Source: NFPA 921, 2001 edition, Figure 17.3.1(c).

EXAMPLE FORM FOR DATA FOR COMPARTMENT FIRE MODELING

Room Number _____ Use _____

Size (use diagrams if possible)   Wall /floor /ceiling

Construction

   Length _____   _____

   Width _____   _____

   Height _____   _____

Lining Materials (that represent over 10% of room lining)
(Include thickness, density, and other material characteristics if known)

   Wall material   Percentage of walls or area involved

   _____   _____

   _____   _____

   _____   _____

   Ceiling material

   _____   _____

   _____   _____

   Floor or floor covering material

   _____   _____

   _____   _____

Doors, windows, and other openings [Enter all heights as distance above floor. If door sill is at floor, enter zero (0).]

Openings	to top	to sill	Width	Changes during fire (how?)[1]
_____	_____	_____	_____	_____
_____	_____	_____	_____	_____
_____	_____	_____	_____	_____
_____	_____	_____	_____	_____
_____	_____	_____	_____	_____
_____	_____	_____	_____	_____

[1] For example: "Window broke at 10:33" or "Door was closed until opened by escaping occupant, then left open — Exit Time 10:30."

(Page 1 of 2)

Figure 2.75 Data collection sheet for compartment fire modeling.
Source: NFPA 921, 2001 edition, Figure A.17.5.2.

DATA FOR COMPARTMENT FIRE MODELING

Heating, ventilation, and air conditioning (HVAC). Include air flows from HVAC systems. Give rates and positions of, supply and return or exhaust in this room. Also sizes and types of ducts/diffusers.

_____

_____

_____

Tightness of walls, closed windows, door fits, etc. (Unless fit is very loose, classify as tight, average, or loose. If fit is very loose, try to get size, number, and location of cracks, etc.)

Doors _____

Windows _____

Inside Walls _____

Exterior Walls _____

Fire history (List all significant events involving progress of the fire.)

Time (hard or soft)	Event
e.g. 1:10 am	sofa involved, flames 3 feet high
1:17 am	room flashover
1:19 am	large fire plume into hallway
1:23 am	smoke out of third floor window

Initial fuel item(s) description

Description	Size	Material
e.g. sofa	full	polyurethane, with cotton upholstery

Suspected ignitor (List ignitor if known with qualification on confidence.)

Ignitor: _e.g. cigarette_ _____

Confidence: _probable_ _____

(Page 2 of 2)

Figure 2.75 *(continued)*

FIRE SCENE DOCUMENTATION AND ANALYSIS
Analytical Tools for the Fire Investigator

2-67

Table 2.3 Sample Failure Mode and Effects Tabulation for Lunchroom Fire

Component Item	Failure Mode	Cause of Failure	Effects of Failure	Hazard Created	Necessary Conditions	Indication of Failure
Coffee maker	Heater current flows without shutoff	Switch left on and controls fail	"Boils" out any water in reservoir Thermal runaway of heating element Local temperature increases above 600°C	Ignition of plastic housing	Power on Switch on or fails closed Thermostat fails in "on" position Both thermal fuses fail to open	Melting of aluminum housing around heating element Condensed aluminum at base of maker Thermostat closed circuit Both fuses closed circuit
Range (electric)	Autoignition of cooking oil	Unattended cooking Control failure	Oil temperature raised above autoignition temperature	Burning oil fire and large amount of smoke	Unit on Switch on or fails in closed position No temperature regulation	Burner control in on position Melted aluminum pan Oil consumed or spilled on unit Contacts fused or welded

Note: The data and conclusions presented in this table are hypothetical and used for example purposes only.
Source: NFPA 921, 2001 edition, Table 17.3.2.

PART 3

BUILDING CONSTRUCTION AND SYSTEMS

Part 3 of the *Field Guide for Fire Investigators* is intended to assist the fire investigator in identifying components of building construction and building systems. This Part consists of the following elements:

- Building Construction
- Electrical Systems
- Fire Protection Systems

BUILDING CONSTRUCTION

Proper identification and description of building construction and structure at the scene is important both to assist investigators in preparing reports and to provide an accurate description for the use of others who may be called to rely upon their results. The following materials will provide the investigator with figures and diagrams to aid in identification of general types of construction and construction materials and to assist in the identification of individual components or structural elements.

Figures 3.1 through 3.21 illustrate types of building construction and typical wall, frame, and flooring assemblies.

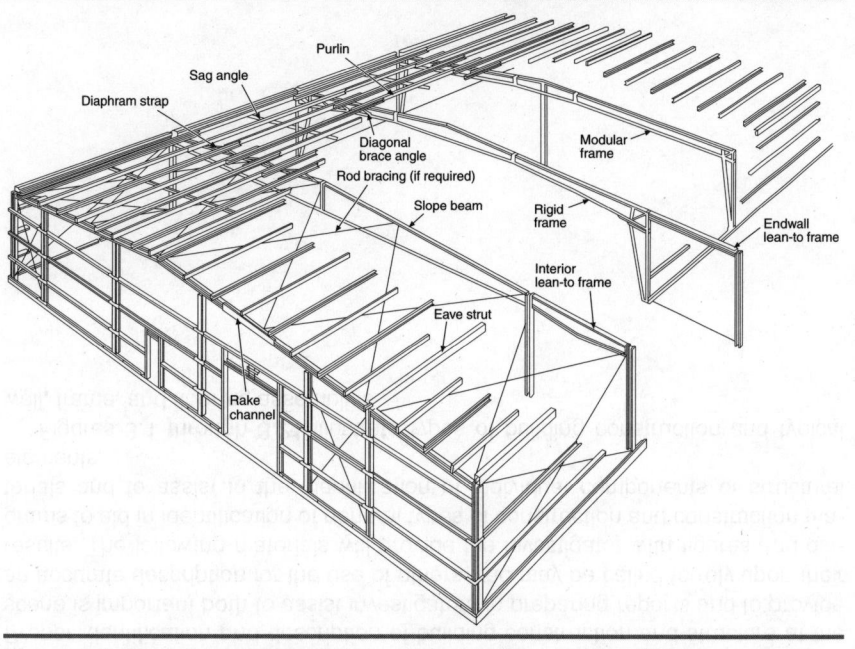

Figure 3.1 Framing representative of Type II (noncombustible) construction.
Source: *Fire Protection Handbook,* 19th edition, Figure 12.2.1.

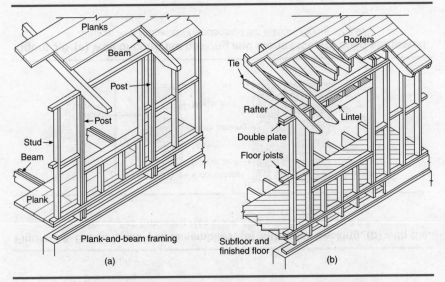

Figure 3.2 Two variations on basic Type V construction.
Source: *Fire Protection Handbook,* 19th edition, Figure 12.2.4.

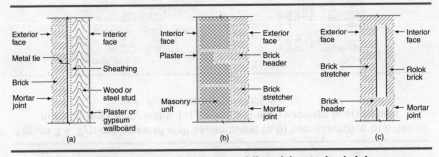

Figure 3.3 Typical types of wall assemblies: (a) exterior brick veneer on wood-frame wall, (b) 12-in. (305-mm) exterior faced or veneered wall, (c) 8-in. (203-mm) exterior hollow Rolok Bak®.
Source: *Fire Protection Handbook,* 19th edition, Figure 12.2.6.

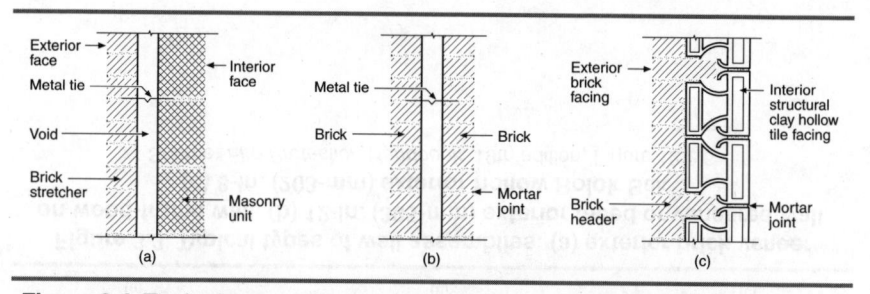

Figure 3.4 Typical types of wall assemblies: (a,b) two examples of exterior nonbearing cavity walls, (c) 12-in. (305-mm) exterior bearing wall.
Source: *Fire Protection Handbook,* Figure 12.2.7.

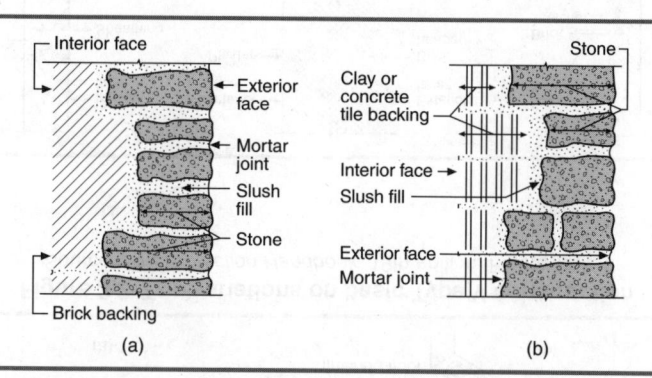

Figure 3.5 Stone-faced wall assemblies: (a) with brick backing, (b) with tile.
Source: *Fire Protection Handbook,* 19th edition, Figure 12.2.8.

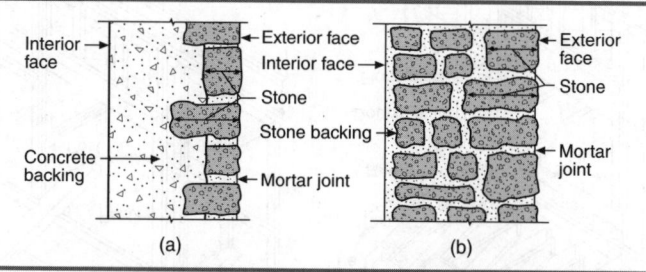

Figure 3.6 (a) Wall of concrete backing and stone face, (b) solid stone wall.
Source: *Fire Protection Handbook,* 19th edition, Figure 12.2.9.

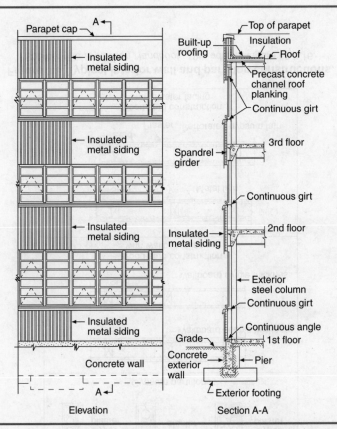

Figure 3.7 Elevation of typical bay of steel frame curtain wall construction.
Source: *Fire Protection Handbook,* 19th edition, Figure 12.2.13.

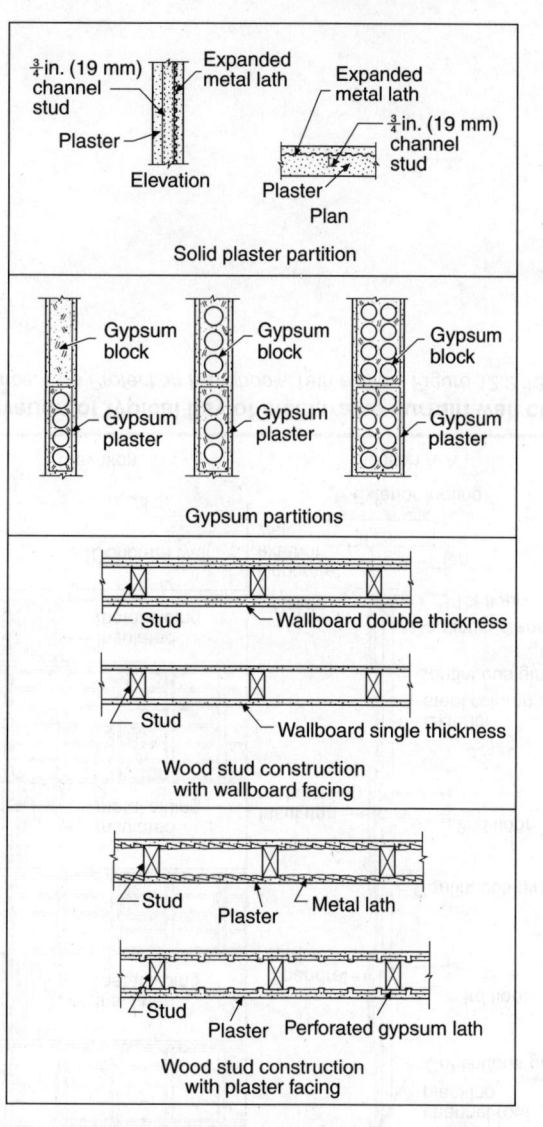

Figure 3.8 Typical interior wall and partition constructions.
Source: *Fire Protection Handbook,* 19th edition, Figure 12.2.16.

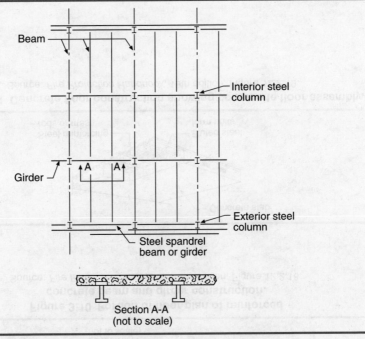

Figure 3.9 Portion of floor plan of steel frame structure.
Source: *Fire Protection Handbook,* 19th edition, Figure 12.2.17.

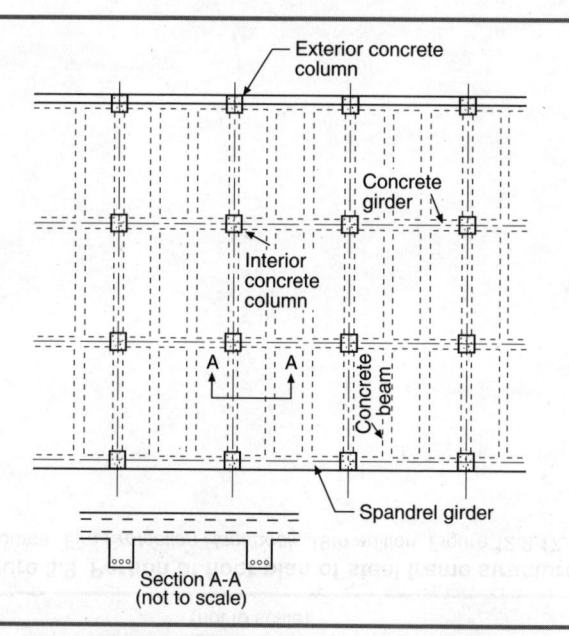

**Figure 3.10 Portion of floor plan of reinforced
concrete beam and girder construction.**
Source: *Fire Protection Handbook,* 19th edition, Figure 12.2.18.

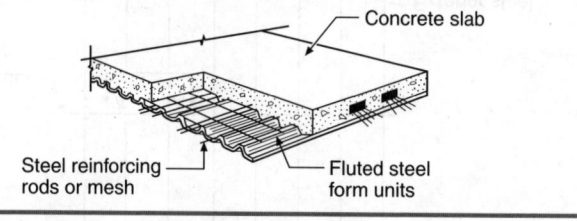

Figure 3.11 Concrete floor construction showing composite floor assembly.
Source: *Fire Protection Handbook,* 19th edition, Figure 12.2.19.

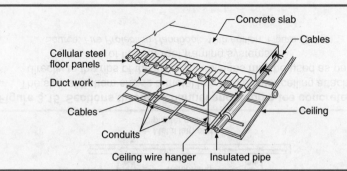

Figure 3.12 Concrete floor poured over light gauge cellular floor panels.
Source: *Fire Protection Handbook,* 19th edition, Figure 12.2.20.

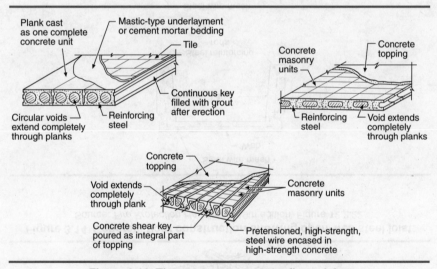

Figure 3.13 Three precast concrete floor slabs.
Source: *Fire Protection Handbook,* 19th edition, Figure 12.2.21.

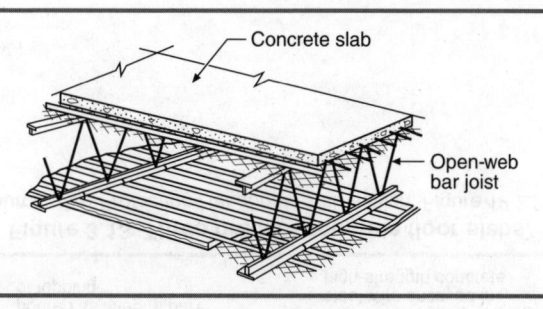

Figure 3.14 Concrete floor construction showing open-web steel joist.
Source: *Fire Protection Handbook,* 19th edition, Figure 12.2.22.

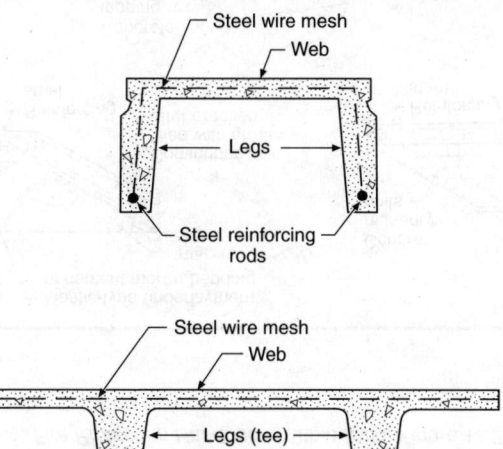

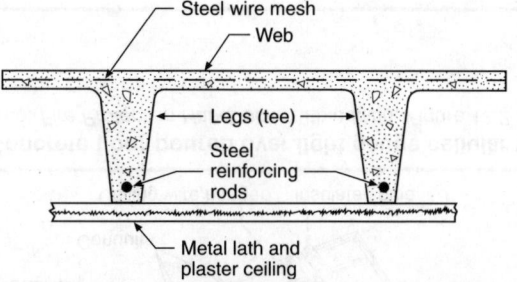

Figure 3.15 Sections through channel and double-tee concrete slabs.
The slab at bottom shows a metal lath and plaster ceiling attached
directly to the ribs of the slab. These slabs may be used as part
of floor or roof framing systems.
Source: *Fire Protection Handbook,* 19th edition, Figure 12.2.24.

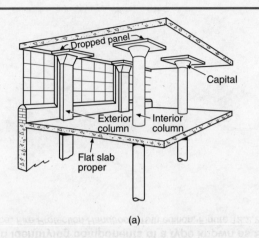

(a)

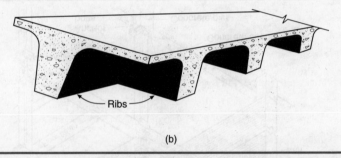

(b)

**Figure 3.16 Two types of two-way flat slabs: (a) flat underside,
(b) two-way ribbed underside.**
Source: *Fire Protection Handbook,* 19th edition, Figure 12.2.25.

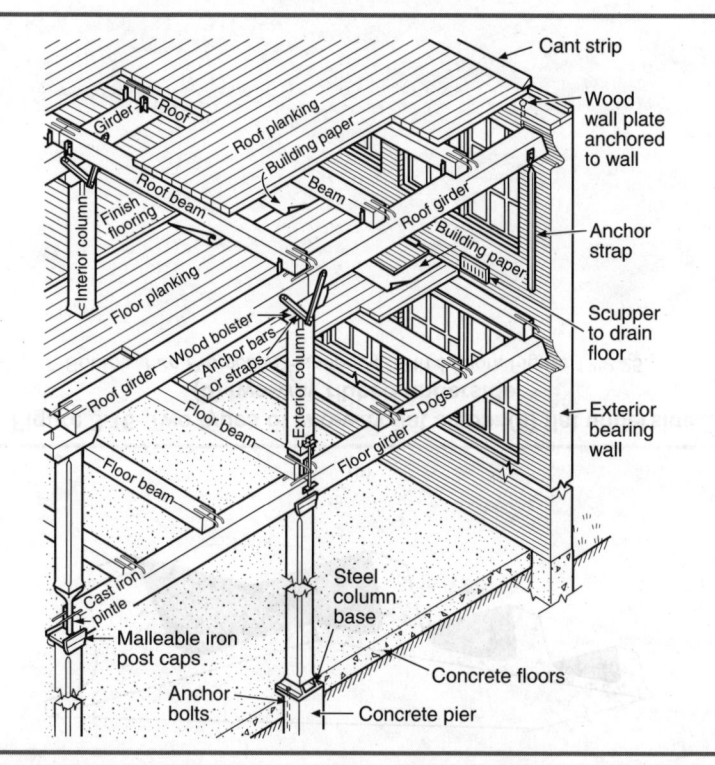

Figure 3.17 Components of a heavy timber building showing floor framing and identifying components of a type known as Semimill.
Source: *Fire Protection Handbook,* 19th edition, Figure 12.2.27.

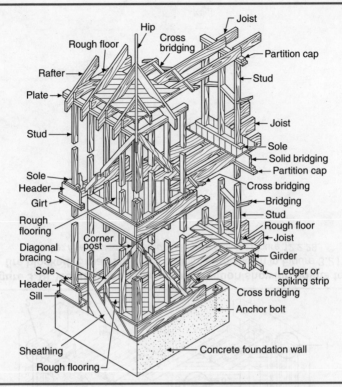

Figure 3.18 Wood-frame platform construction common to dwellings with structural members identified.
Source: *Fire Protection Handbook,* 19th edition, Figure 12.2.28.

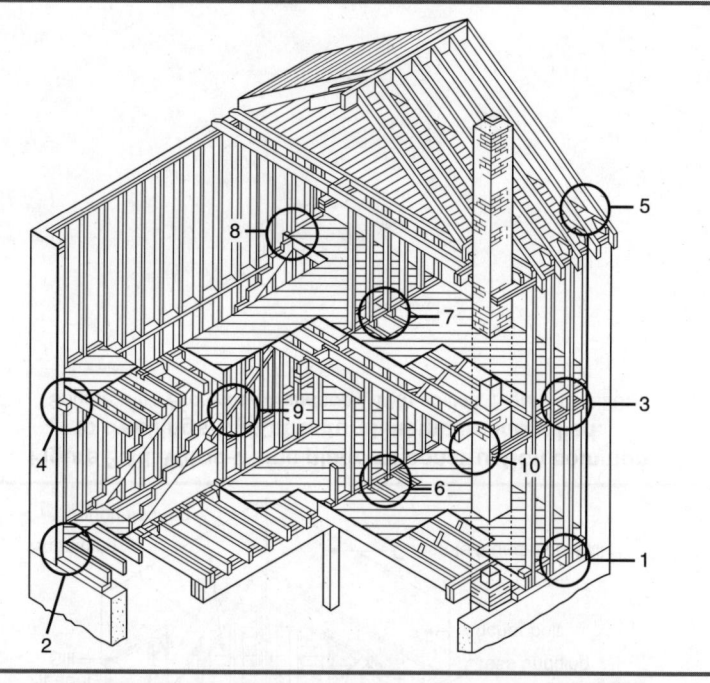

Figure 3.19 Wood balloon frame construction showing points to be fire blocked. Expanded views of points are shown in Figure 3.21.
Source: *Fire Protection Handbook,* 19th edition, Figure 12.2.29.

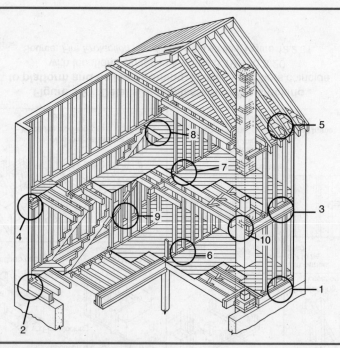

Figure 3.20 Wood platform frame construction showing points to be fire blocked. Expanded views of points are shown in Figure 3.21.
Source: *Fire Protection Handbook,* 19th edition, Figure 12.2.30.

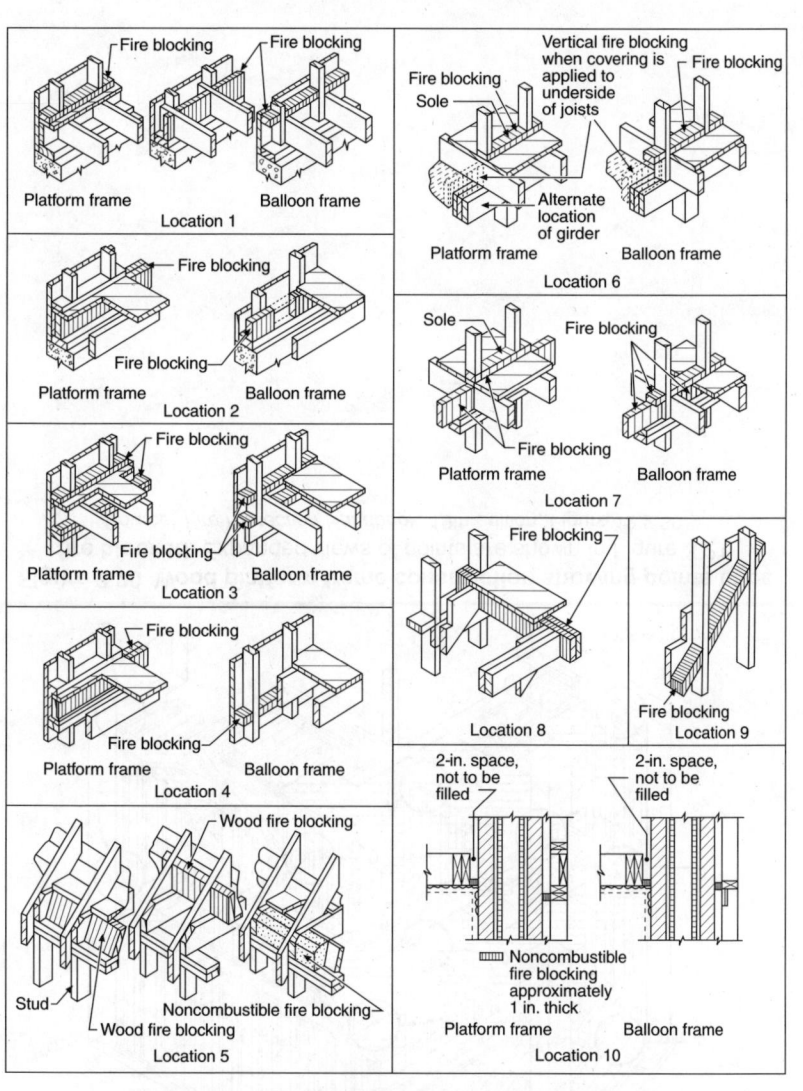

**Figure 3.21 Details of application of fire blocking
to platform and balloon framing.** Location numbers coincide
with locations circled in Figures 3.19 and 3.20.
Source: *Fire Protection Handbook,* 19th edition, Figure 12.2.31.

Roof construction and materials can play a role in fire origin, spread, and growth investigations. Ignition of combustible materials may be the start of a fire or may result in spread to other portions of a building. The type of construction may be a factor in venting of a fire or early collapse. Diagrammatic examples of some floor systems are provided below. More information and selected references can be found in the NFPA *Fire Protection Handbook*.

Table 3.1 and Figures 3.22, 3.23, and 3.24 illustrate typical roof assemblies.

Table 3.1 Some Typical Prepared Roof Coverings

Roof Covering[a]	Minimum/ Maximum Roof Slope, in. per ft	Class A	Class B	Class C
Brick Concrete Tile Slate	minimum required for drainage/ unlimited	Brick, 2-1/4 in. thick. Reinforced Portland cement, 1 in. thick. Concrete or clay floor or deck tile, 1 in. thick. Flat or French-type clay or concrete tile, 3/8 in. thick with 1-1/2 in. or more end lap[b] and head lock, spacing body of tile 1/2 in. or more above roof sheathing, with underlay of one layer of Type 30 or two layers of Type 15 asphalt-saturated organic felt. Clay or concrete roof tile, Spanish or Mission pattern, 7/16 in. thick, 3 in. end lap,[b] same underlay as above. Slate, 3/16 in. thick, laid American method.		

Continued

Table 3.1 Some Typical Prepared Roof Coverings *(continued)*

Roof Covering[a]	Minimum/ Maximum Roof Slope, in. per ft	Class A	Class B	Class C
Metal roofing	12/unlimited	Sheet roofing of 16-oz copper or of 30-gauge steel or iron protected against corrosion. Limited to noncombustible roof decks or noncombustible roof supports when no separate roof deck is provided.	Sheet roofing of 16-oz copper or of 30-gauge steel or iron protected against corrosion or shingle-pattern roofings with underlay of one layer of Type 30 or two layers of Type 15 asphalt-saturated organic felt.[b]	Sheet roofing of 16-oz copper or of 30-gauge steel or iron, protected against corrosion, or shingle-pattern roofings, either without underlay or with underlay of rosin-sized paper.
	1/4 (standing seam) or 1/2 (lap seam)/ unlimited	Standing or lap seam metal panel with glass fiber batts below on steel purling or joints.		Zinc sheets or shingle roofings with an underlay of one layer of Type 30 or two layers of Type 15 asphalt-saturated organic felt.[b]
Organic-felt (previously referred to as rag felt) sheet coverings	1/4/unlimited			Sheet coverings of asphalt organic felt either grit surfaced or aluminum surfaced.
Organic-felt (previously referred to as rag felt) shingle covering, with special coating	Sufficient to permit drainage[c]	Mineral surfaced, two or more thicknesses.	Mineral surfaced, two or more thicknesses.	
Organic-felt (previously referred to as rag felt) shingle coverings	Sufficient to permit drainage[c]	Mineral surfaced, two or more thicknesses.	Mineral surfaced, two or more thicknesses.	Mineral surfaced shingles, one or more thicknesses.
Asphalt glass fiber mat shingle coverings	Sufficient to permit drainage[c]	Mineral surfaced, two or more thicknesses.	Mineral surfaced shingles, one or more thicknesses.	Mineral surfaced shingles, one or more thicknesses.
Asphalt glass mat sheet covering	Sufficient to permit drainage[c]			Mineral surfaced.

Table continues below.

Table 3.1 *(continued)*

Roof Covering[a]	Minimum/ Maximum Roof Slope, in. per ft	Class A	Class B	Class C
Fire-retardant treated red cedar wood shingles and shakes				Treated shingles or shakes, one or more thicknesses; shakes require at least one layer of Type 15 felt underlayment.

Note: Roofing materials containing asbestos are no longer used in the United States; however, they may still be encountered in existing construction.

[a]Prepared roof coverings as classified as applied over square-edge wood sheathing of 1-in. nominal thickness, or the equivalent, unless otherwise specified. See footnotes (build-up roof coverings) to Table 12.2.9 in the *Fire Protection Handbook*. Laid in accordance with instruction sheets accompanying package. Limited to decks capable of receiving and retaining nails. Prepared roofings are labeled by Underwriters' Laboratories to indicate the classification when applied in accordance with direction for application included in package.

[b]End lap means the overlapping length of the two units, one placed over the other. Head lap in shingle-type roofs is the distance a shingle in any course overlaps a shingle in the second course below it. However, with shingles laid by the Dutch-lap method, where no shingle overlaps a shingle in the second course below, the head lap is taken as the distance a shingle overlaps one in the next course below.

[c]Typically 1/4 in. per ft

For SI units: 1 in. = 25.4 mm; 1 ft = 0.305 m; 1 oz = 0.0284 kg; 1 lb = 0.454 kg.

Source: *Fire Protection Handbook*, Table 12.2.8.

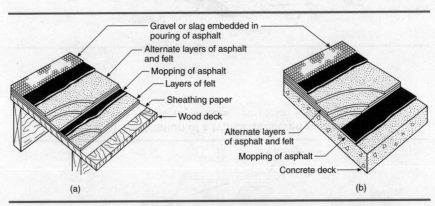

Gravel or slag embedded in pouring of asphalt
Alternate layers of asphalt and felt
Mopping of asphalt
Layers of felt
Sheathing paper
Wood deck

Alternate layers of asphalt and felt
Mopping of asphalt
Concrete deck

(a)
(b)

Figure 3.22 Typical built-up roof coverings with hot asphalt mopping or coal tar pitch between each two layers: (a) 5-ply built-up roof over wood, (b) 4-ply built-up roof over concrete.
Source: *Fire Protection Handbook*, 19th edition, Figure 12.2.36.

BUILDING CONSTRUCTION AND SYSTEMS
Building Construction

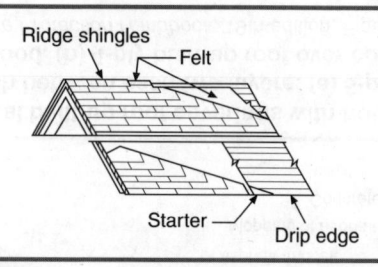

Figure 3.23 Installation of a typical prepared roof covering.
Source: *Fire Protection Handbook,* 19th edition, Figure 12.2.37.

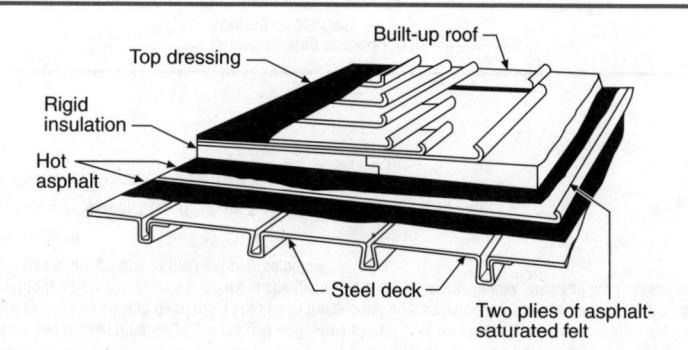

Figure 3.24 Typical built-up roof covering with a combustible vapor barrier adhered to the roof deck and to roof insulation by a combustible adhesive.
Source: *Fire Protection Handbook,* 19th edition, Figure 12.2.38.

ELECTRICAL SYSTEMS

Proper identification and description of a building's electrical supply, distribution, branch circuit opponents, and overcurrent protection is important to assist investigators in identifying or eliminating possible ignition scenarios, preparing reports, and providing an accurate description for the use of others who may be called to rely upon their results.

Electrical System Components

Figures 3.25 through 3.43 provide the investigator with figures and diagrams to aid in identification of electrical components for buildings with 120/240-volt single-phase electrical systems.

	Solid Copper Wire	
	Diameter	Resistance in ohms per 1000 ft (305 m) at 158°F (70°C)
14 AWG —— ● ——	.064 in. (1.63 mm)	3.1
12 AWG —— ● ——	.081 in. (2.06 mm)	2.0
10 AWG —— ● ——	.102 in. (2.60 mm)	1.2

Figure 3.25 Sizes, cross sections, and resistance of solid copper wire.
Source: NFPA 921, 2001 edition, Figure 6.2.10.

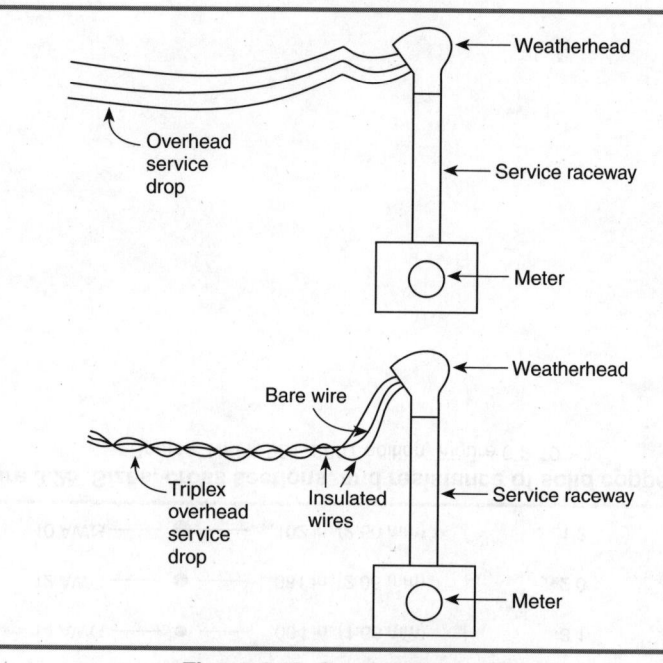

Figure 3.26 Overhead service.
Source: NFPA 921, 2001 edition, Figure 6.3.2.1(b).

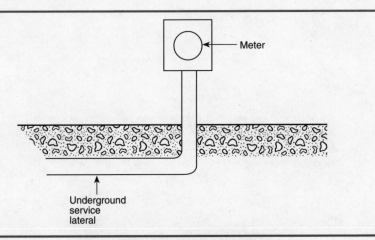

Figure 3.27 Underground service.
Source: NFPA 921, 2001 edition, Figure 6.3.2.1(c).

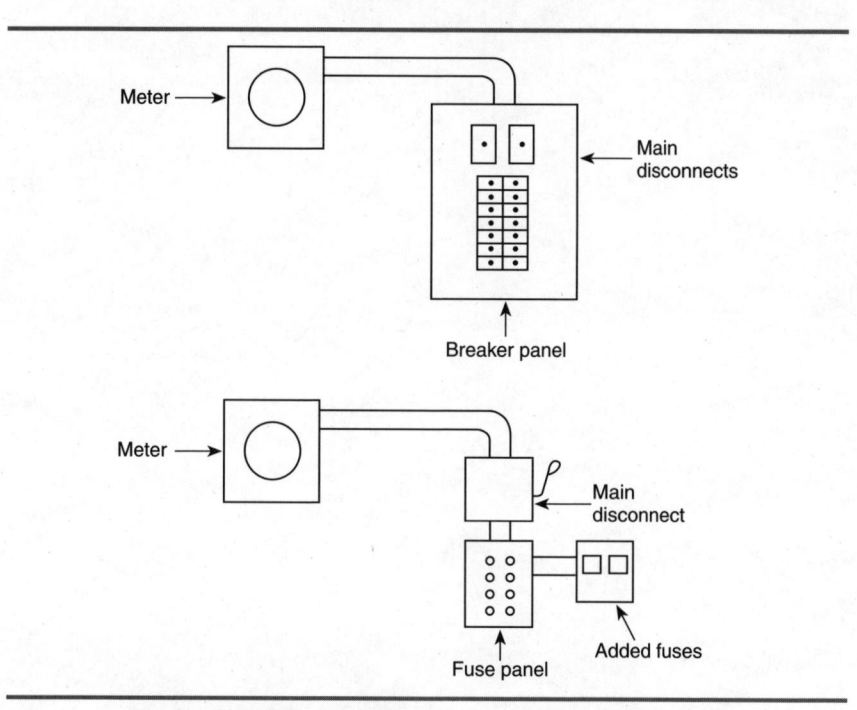

Figure 3.28 Service entrance and service equipment.
Source: NFPA 921, 2001 edition, Figure 6.3.3.

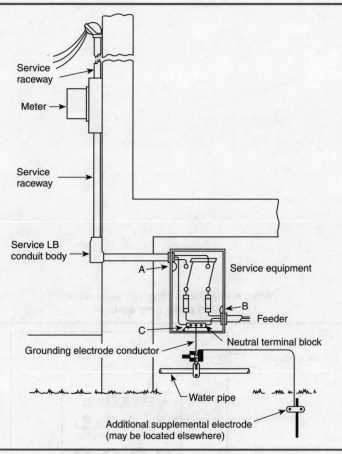

Service raceway

Meter

Service raceway

Service LB conduit body

A

Service equipment

B

Feeder

C

Neutral terminal block

Grounding electrode conductor

Water pipe

Additional supplemental electrode
(may be located elsewhere)

Figure 3.29 Grounding at a typical small service.
Source: NFPA 921, 2001 edition, Figure 6.5.

BUILDING CONSTRUCTION AND SYSTEMS
Electrical Systems

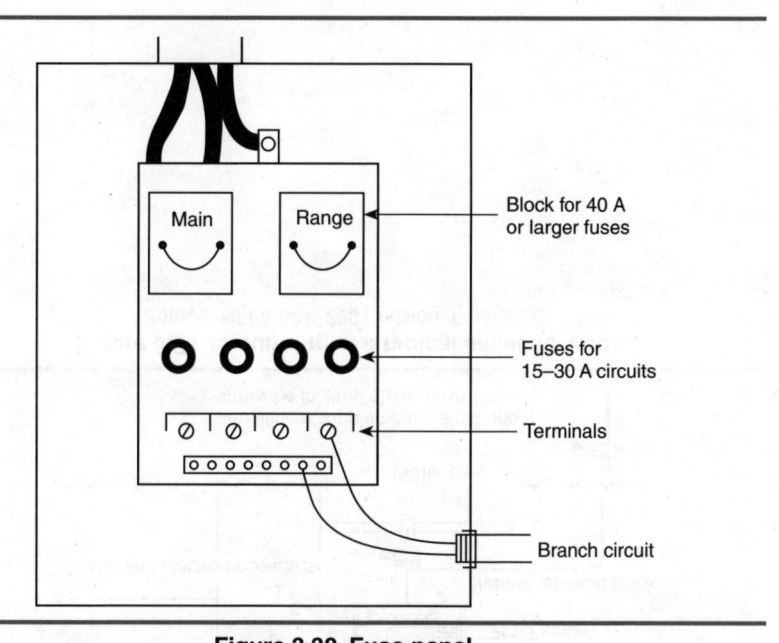

Figure 3.30 Fuse panel.
Source: NFPA 921, 2001 edition, Figure 6.6(a).

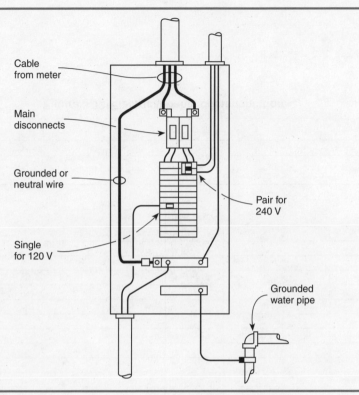

Cable from meter

Main disconnects

Grounded or neutral wire

Single for 120 V

Pair for 240 V

Grounded water pipe

Figure 3.31 Common arrangement for a circuit breaker panel.
Source: NFPA 921, 2001 edition, Figure 6.6(b).

ELECTRICAL PANEL DOCUMENTATION

Fire Location:	Date:	Case #:

Panel Location:	Main Size:	Fuses: ❑
		Circuit Breakers: ❑

		LEFT BANK				RIGHT BANK	
#	Rating Amps	Labeled Circuit	Status	#	Rating Amps	Labeled Circuit	Status
1	—		—	2	—		—
3	—		—	4	—		—
5	—		—	6	—		—
7	—		—	8	—		—
9	—		—	10	—		—
11	—		—	12	—		—
13	—		—	14	—		—
15	—		—	16	—		—
17	—		—	18	—		—
19	—		—	20	—		—
21	—		—	22	—		—
23	—		—	24	—		—
25	—		—	26	—		—
27	—		—	28	—		—
29	—		—	30	—		—
Notes:				Notes:			

Documented by:

Figure 3.32 Electrical panel documentation.

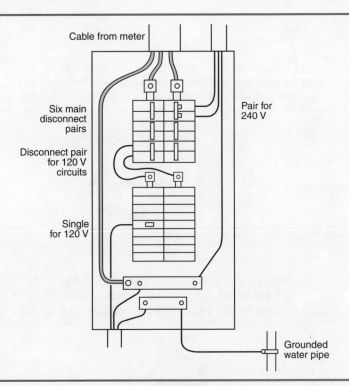

Cable from meter

Six main
disconnect
pairs

Disconnect pair
for 120 V
circuits

Single
for 120 V

Pair for
240 V

Grounded
water pipe

Figure 3.33 Common arrangement for a split-bus circuit-breaker panel.
Source: NFPA 921, 2001 edition, Figure 6.6(c).

BUILDING CONSTRUCTION AND SYSTEMS
Electrical Systems

**Figure 3.34 A typical Edison-based nonrenewable fuse,
single element, for replacement purposes only.**
Source: NFPA 921, 2001 edition, Figure 6.6.1(a).

**Figure 3.35 Another Edison-based non-renewable fuse,
dual element, for replacement purposes only.**
Source: NFPA 921, 2001 edition, Figure 6.6.1(b).

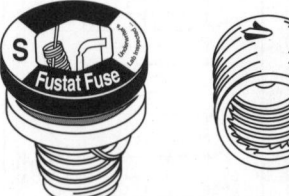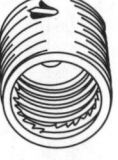

**Figure 3.36 A Type S non-renewable fuse and adapter.
The time-lag type of fuse is acceptable but not required.**
Source: NFPA 921, 2001 edition, Figure 6.6.1(c).

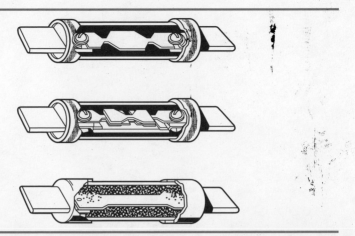

Figure 3.37 Typical cartridge fuses. Top, an ordinary drop-out link renewable fuse; center, a super lag renewable fuse; and bottom, a one-time fuse.
Source: NFPA 921, 2001 edition, Figure 6.6.1.4(a).

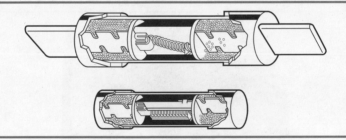

Figure 3.38 Dual-element cartridge fuses, blade and ferrule types.
Source: NFPA 921, 2001 edition, Figure 6.6.1.4(b).

BUILDING CONSTRUCTION AND SYSTEMS
Electrical Systems

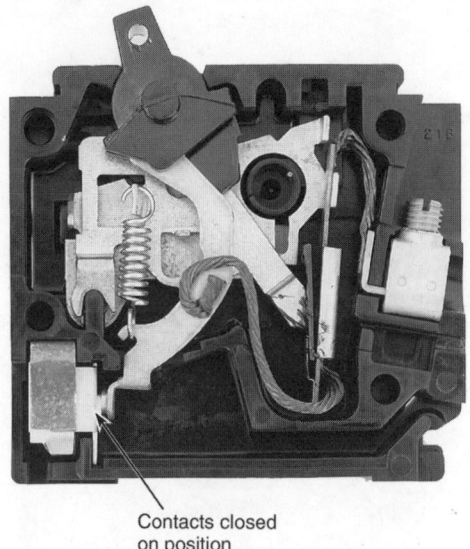

Contacts closed
on position

**Figure 3.39 A 15 A residential-type circuit breaker
in the closed (on) position.**
Source: NFPA 921, 2001 edition, Figure 6.6.2(a).

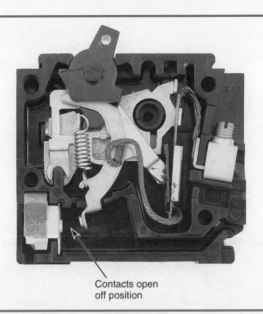

Contacts open
off position

Figure 3.40 A 15 A residential-type circuit breaker in the open (off) position.
Source: NFPA 921, 2001 edition, Figure 6.6.2(b).

BUILDING CONSTRUCTION AND SYSTEMS
Electrical Systems

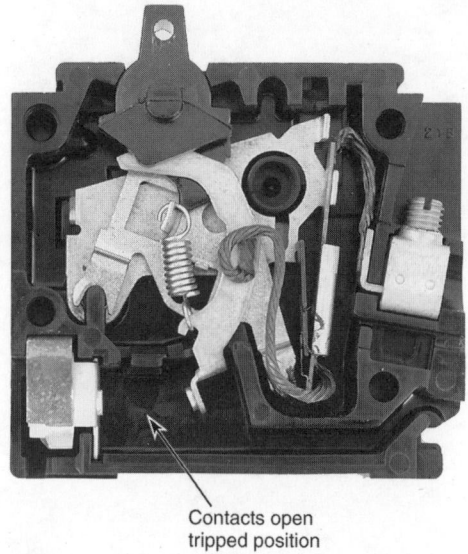

Contacts open
tripped position

**Figure 3.41 A 15 A residential-type circuit breaker
in the open (tripped) position.**
Source: NFPA 921, 2001 edition, Figure 6.6.2(c).

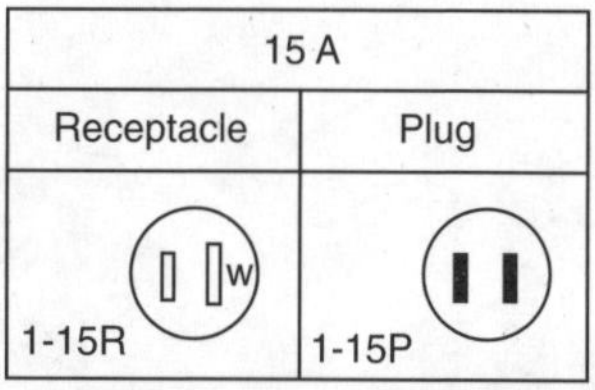

15 A	
Receptacle	Plug
[[]]w 1-15R	(II) 1-15P

Figure 3.42 Nongrounding-type receptacle.
Source: NFPA 921, 2001 edition, Figure 6.8.2(a).

15 A		20 A	
Receptacle	Plug	Receptacle	Plug
5-15R	5-15P	5-20R	5-20P

Figure 3.43 Grounding-type receptacle.
Source: NFPA 921, 2001 edition, Figure 6.8.2(b).

Samples of Fire Damage to Electrical Components

Figures 3.44 through 3.58 provide the investigator with figures and diagrams to aid in identification of conductor damage for buildings with 120/240-volt single-phase electrical systems.

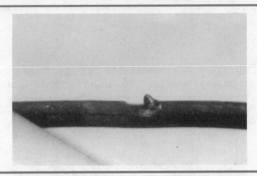

Figure 3.44 A solid copper conductor notched by a short circuit.
Source: NFPA 921, 2001 edition, Figure 6.10.2(a).

BUILDING CONSTRUCTION AND SYSTEMS
Electrical Systems

Figure 3.45 Stranded copper lamp cord that was severed by a short circuit.
Source: NFPA 921, 2001 edition, Figure 6.10.2(b).

Figure 3.46 Copper conductors severed by arcing through the insulation.
Source: NFPA 921, 2001 edition, Figure 6.10.3(a).

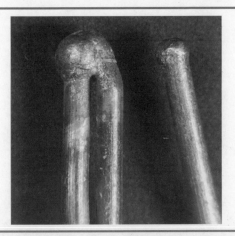

Figure 3.47 Copper conductors severed by arcing through the charred insulation with a large bead welding the two conductors together.
Source: NFPA 921, 2001 edition, Figure 6.10.3(b).

Figure 3.48 Stranded copper conductors severed by arcing through charred insulation with the strands terminated in beads.
Source: NFPA 921, 2001 edition, Figure 6.10.3(c).

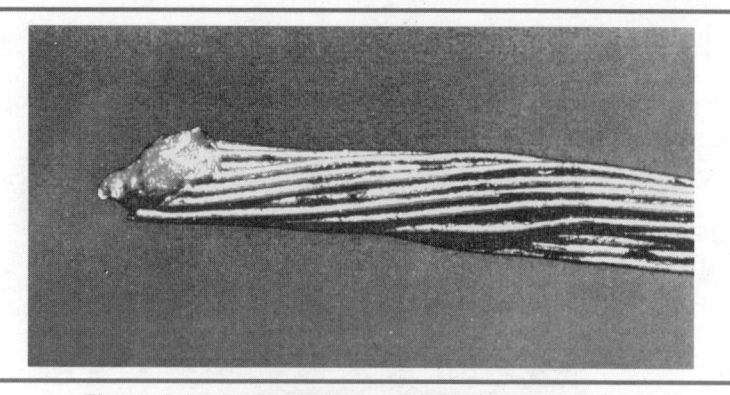

Figure 3.49 Arc damage to 18 AWG cord by arcing through charred insulation.
Source: NFPA 921, 2001 edition, Figure 6.10.3(d).

Figure 3.50 Spot arc damage to 14 AWG conductor caused by arcing through charred insulator (lab test).
Source: NFPA 921, 2001 edition, Figure 6.10.3(e).

BUILDING CONSTRUCTION AND SYSTEMS
Electrical Systems

**Figure 3.51 Arc damage to 18 AWG cord by arcing
through charred insulation (lab test).**
Source: NFPA 921, 2001 edition, Figure 6.10.3(f).

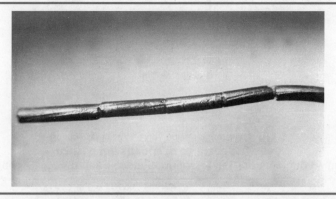

Figure 3.52 Aluminum conductor severed by overcurrent showing offsets.
Source: NFPA 921, 2001 edition, Figure 6.10.5.

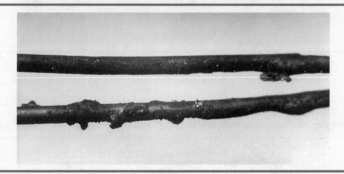

Figure 3.53 Copper conductors fire-heated to the melting temperature, showing regions of flow of copper, blistering, and no surface distortion.
Source: NFPA 921, 2001 edition, Figure 6.10.6.2(a).

Figure 3.54 Fire-heated copper conductors, showing globules.
Source: NFPA 921, 2001 edition, Figure 6.10.6.2(b).

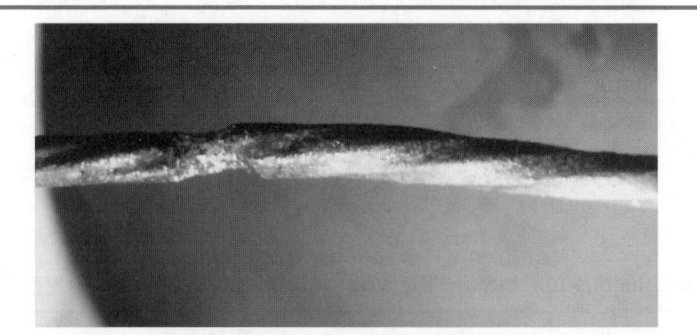

Figure 3.55 Stranded copper conductor in which melting by fire caused the strands to be fused together.
Source: NFPA 921, 2001 edition, Figure 6.10.6.2(c).

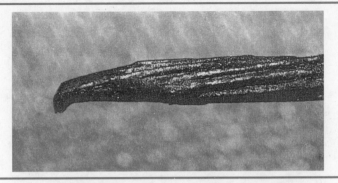

Figure 3.56 Fire melting of stranded copper wire.
Source: NFPA 921, 2001 edition, Figure 6.10.6.2(d).

Figure 3.57 Another example of fire melting of stranded copper wire.
Source: NFPA 921, 2001 edition, Figure 6.10.6.2(e).

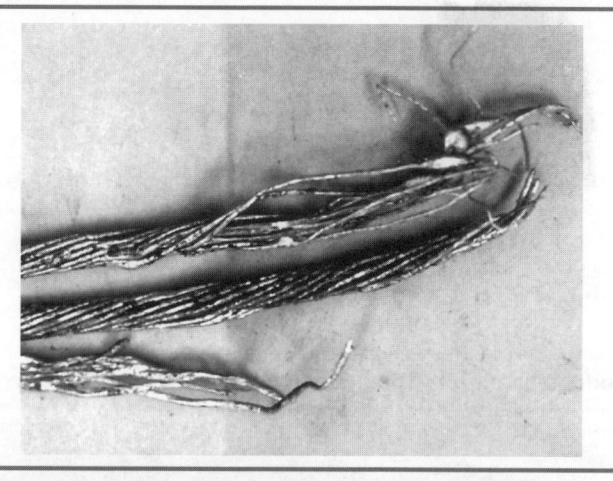

**Figure 3.58 Aluminum cables that were melted by fire,
showing thinned areas, bulbous areas, and pointed ends.**
Source: NFPA 921, 2001 edition, Figure 6.10.6.2(f).

FIRE PROTECTION SYSTEMS

The performance of fire protection systems is an important part of the analysis of how and why a fire spread and in the analysis of the reasons behind the magnitude of a fire loss. Fire protection systems in buildings include fire resistive construction and compartmentation, smoke management, egress systems, and fire suppression and detection systems. In evaluation of the performance of these systems, the initial determination should be whether or not the fire protection features functioned.

There are a number of things to look for in general and questions to ask. For example, did the fire door close? If not, was it blocked open? Did the sprinkler heads open? Which ones operated and which did not? Were the water supply valves on or off during the fire? Did the detection system respond? If it did was the alarm sounded locally or transmitted to the fire department?

The focus of this section will be on identification of the basic types of fire alarm systems and fire suppression equipment and systems. When documenting these systems on the fire scene or in reports, it is important to correctly identify the general types of systems and component parts for the record and to assist other investigators or specialists should additional investigations be carried out.

Fire Detectors and Fire Alarm Systems

There are many types of devices for detecting fire, including heat, smoke, and flame detectors. Each will have different response characteristics that can be helpful in analyzing the fire growth and spread. It is important to recognize the general types present and to document their location(s) with respect to the origin of the fire, the egress path, and the location of occupants. Some examples of typical detectors are provided in Figures 3.59 through 3.65. More details about the different types and operating characteristics of detectors and alarm systems can be found in Section 9, "Detection and Alarm," of the *Fire Protection Handbook* (19th edition).

Figures 3.66 through 3.69 illustrate typical arrangements of fire alarm systems.

Figure 3.59 Typical fixed-temperature (nonrestorable) heat detector.
Source: Kidde-Fenwal, Ashland, MA; photo courtesy of
Mammoth Fire Alarms, Inc., Lowell, MA.

Figure 3.60 Combination rate-of-rise and fixed-temperature heat detector.
Source: Kidde-Fenwal, Ashland, MA; photo courtesy of
Mammoth Fire Alarms, Inc., Lowell, MA.

Figure 3.61 Typical spot-type smoke detector.
Source: System Sensor Corp., St. Charles, IL.

Figure 3.62 Typical smoke alarm (detector with built in alarm).
Source: Gentex, Inc., Zeeland, MI.

Figure 3.63 Typical combination smoke and heat detector.
Source: System Sensor Corp., St. Charles, IL.

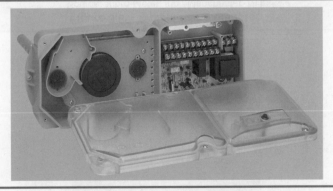

Figure 3.64 Typical duct smoke detector (passive).
Source: System Sensor Corp., St. Charles, IL.

BUILDING CONSTRUCTION AND SYSTEMS
Fire Protection Systems

3-47

Figure 3.65 Typical flame detector.
Source: Vibro-Meter, Inc., Manchester, NH.

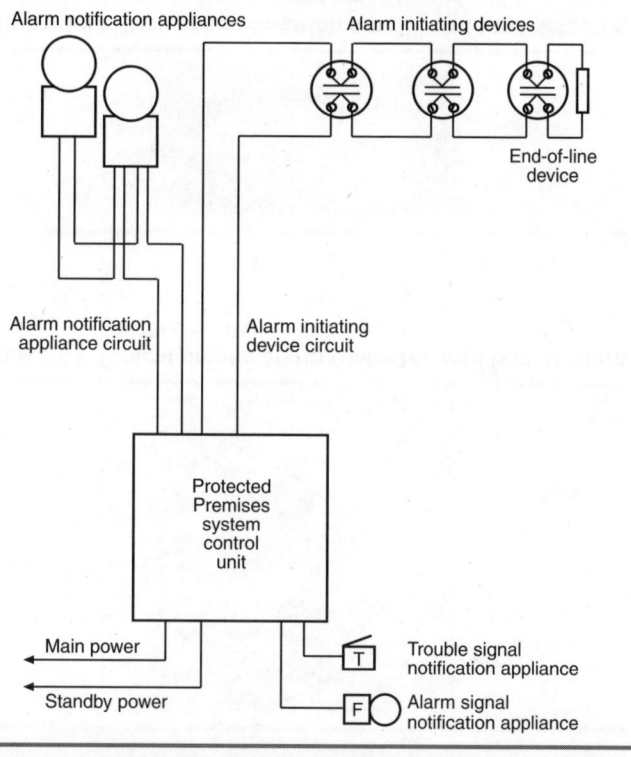

Figure 3.66 Typical arrangement of a local fire alarm system.
Source: *Fire Protection Handbook,* 19th edition, Figure 9.1.2(a).

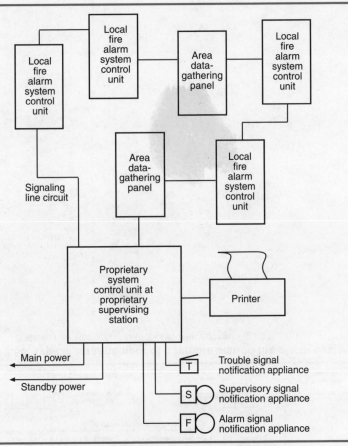

**Figure 3.67 Typical arrangement of a proprietary
protective signaling system.**
Source: *Fire Protection Handbook,* 19th edition, Figure 9.1.4(a).

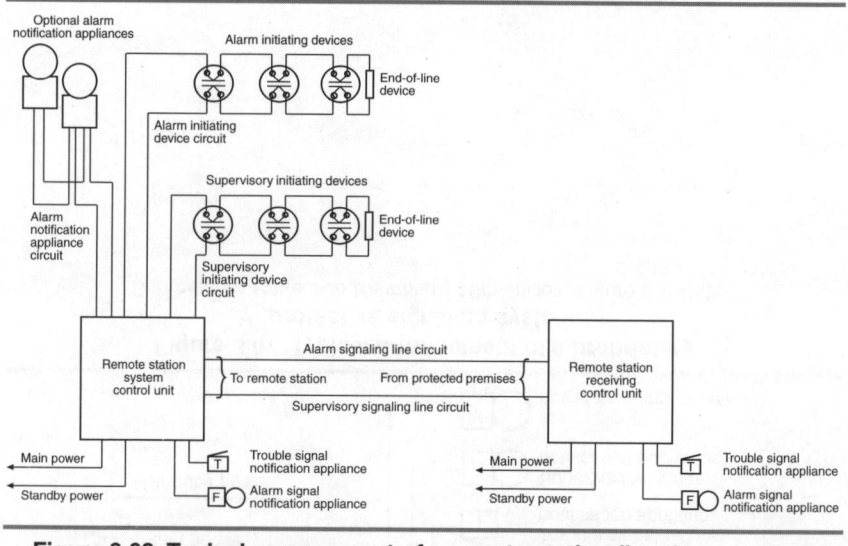

Figure 3.68 Typical arrangement of a remote station fire alarm system.
Source: *Fire Protection Handbook,* 19th edition, Figure 9.1.6(a).

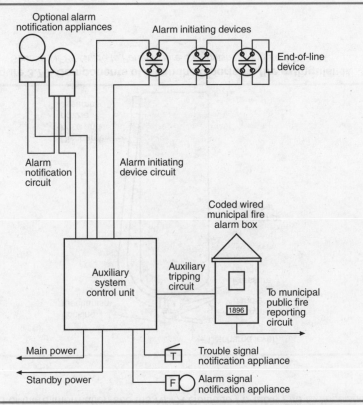

Figure 3.69 Typical arrangement of an auxiliary fire alarm system.
Source: *Fire Protection Handbook,* 19th edition, Figure 9.1.7(a).

Fire Suppression Systems

In examining fire scenes where fire suppression systems and equipment are employed it is important to recognize the types present and their major component parts. This section will cover portable extinguishers and automatic sprinkler systems. In addition to automatic sprinklers, there are a number of specialized fixed fire suppression systems such as CO_2, foam, water mist, dry or wet chemical as well as HALON and its replacement agents. Detailed information about the specialized suppression systems can be found in Section 11, "Fire Suppression without Water," of the NFPA *Fire Protection Handbook* (19th edition).

Portable Fire Extinguishers. In documenting portable extinguishers and their role in a fire incident, the extinguishing agent and the extinguisher's location should be identified as well as whether the unit was discharged. The latter may be determined by reading a gauge or it may require weighing. Figures 3.70 through 3.82 are helpful aids and present the details of the types associated with different agents. For more information see the *NFPA Guide to Portable Fire Extinguishers.*

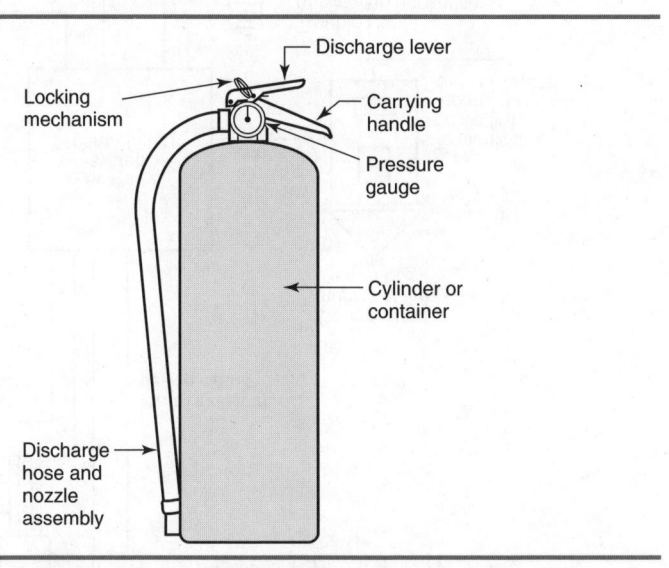

Figure 3.70 Components of a handheld portable fire extinguisher.
Source: *NFPA Guide to Portable Fire Extinguishers,* Figure 1-9, p. 34.

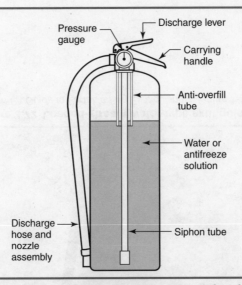

Figure 3.71 Stored-pressure water-type extinguisher.
Source: *NFPA Guide to Portable Fire Extinguishers,* Figure 1-18, p. 43.

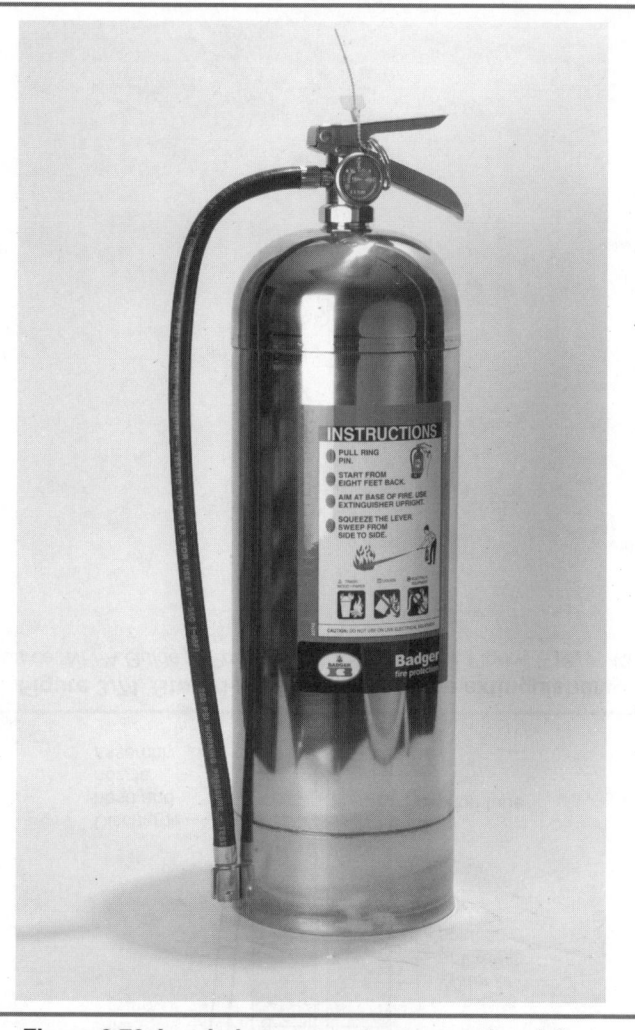

Figure 3.72 Loaded-stream water-type extinguisher.
Source: *NFPA Guide to Portable Fire Extinguishers,* Figure 1-19, p. 45.

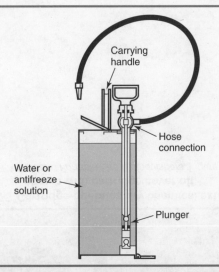

Figure 3.73 Pump tank water-type extinguisher.
Source: *NFPA Guide to Portable Fire Extinguishers,* Figure 1-20, p. 46.

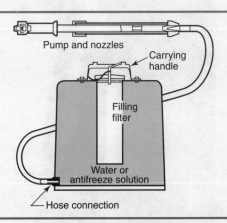

Figure 3.74 Backpack water-type extinguisher.
Source: *NFPA Guide to Portable Fire Extinguishers,* Figure 1-21, p. 47.

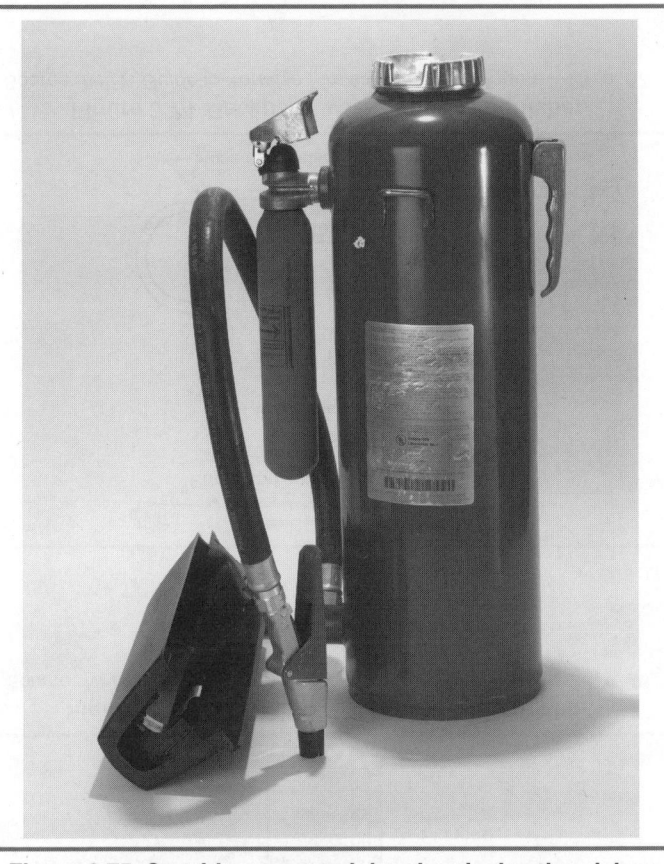

**Figure 3.75 Cartridge-operated dry chemical extinguisher,
shown with cartridge cover off.**
Source: *NFPA Guide to Portable Fire Extinguishers,* Figure 1-22, p. 49.

Figure 3.76 Three sizes of handheld ordinary dry chemical extinguishers.
Source: *NFPA Guide to Portable Fire Extinguishers,* Figure 1-23, p. 50.

Figure 3.77 Three sizes of carbon dioxide fire extinguishers.
Source: *NFPA Guide to Portable Fire Extinguishers,* Figure 1-25, p. 53.

Figure 3.78 Foam-type extinguisher.
Source: *NFPA Guide to Portable Fire Extinguishers,* Figure 1-27, p. 55.

Figure 3.79 Wet chemical extinguisher.
Source: *NFPA Guide to Portable Fire Extinguishers,* Figure 1-28, p. 56.

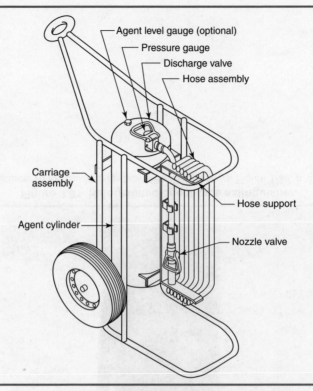

Agent level gauge (optional)
Pressure gauge
Discharge valve
Hose assembly
Carriage assembly
Hose support
Agent cylinder
Nozzle valve

Figure 3.80 Stored-pressure halogenated-agent-type wheeled fire extinguisher.

Source: *NFPA Guide to Portable Fire Extinguishers,* Figure 1-17, p. 42.

Figure 3.81 Halogenated-agent-type extinguisher.
Source: *NFPA Guide to Portable Fire Extinguishers,* Figure 1-29, p. 57.

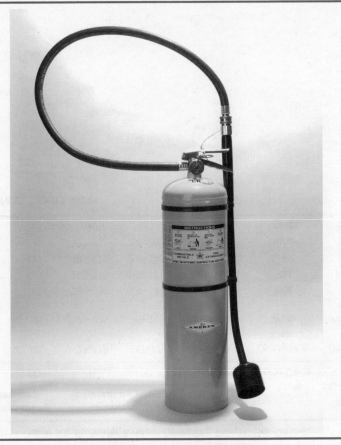

Figure 3.82 Dry powder extinguisher.
Source: *NFPA Guide to Portable Fire Extinguishers,* Figure 1-30, p. 59.

Automatic Sprinkler Systems. Automatic sprinklers are the most common suppression systems and are in wide use in public buildings such as schools, hospitals, libraries, shopping centers, offices, and hotels as well as many residential and industrial buildings. The types of systems can be different depending on the purpose. Among the important information to document is the type of system (wet, dry, pre-action or deluge), status of water supply valves (open or closed), temperature rating of sprinkler heads used, and any obvious factors that might have prevented successful operation such as piping broken by an explosion or items blocking coverage of the water spray.

Table 3.2 and Figures 3.83 through 3.90 are provided to assist in recognizing systems and components for fire scene documentation and reports.

Table 3.2 Sprinkler System Operating Temp Color Codes

Maximum Ceiling Temperature		Temperature Rating		Temperature Classification	Color Code	Glass Bulb Colors
°F	°C	°F	°C			
100	38	135–170	57–77	Ordinary	Uncolored or black	Orange or red
150	66	175–225	79–107	Intermediate	White	Yellow or green
225	107	250–300	121–149	High	Blue	Blue
300	149	325–375	163–191	Extra high	Red	Purple
375	191	400–475	204–246	Very extra high	Green	Black
475	246	500–575	260–302	Ultra high	Orange	Black
625	329	650	343	Ultra high	Orange	Black

Source: NFPA 13, Table 6.2.5.1.

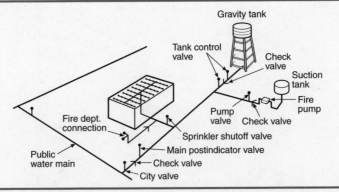

Figure 3.83 Hypothetical sprinkler system installation illustrating various water supply sources and system attachments.

Source: *Fire Protection Handbook,* 19th edition, Figure 10.11.7.

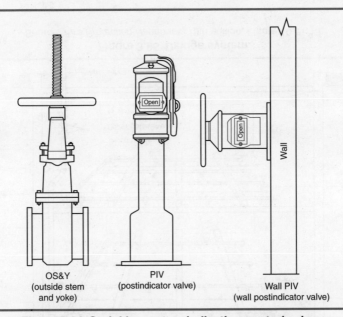

OS&Y
(outside stem
and yoke)

PIV
(postindicator valve)

Wall PIV
(wall postindicator valve)

Figure 3.84 Sprinkler system indicating control valves.

Source: *Fire Protection Handbook,* 19th edition, Figure 10.11.16.

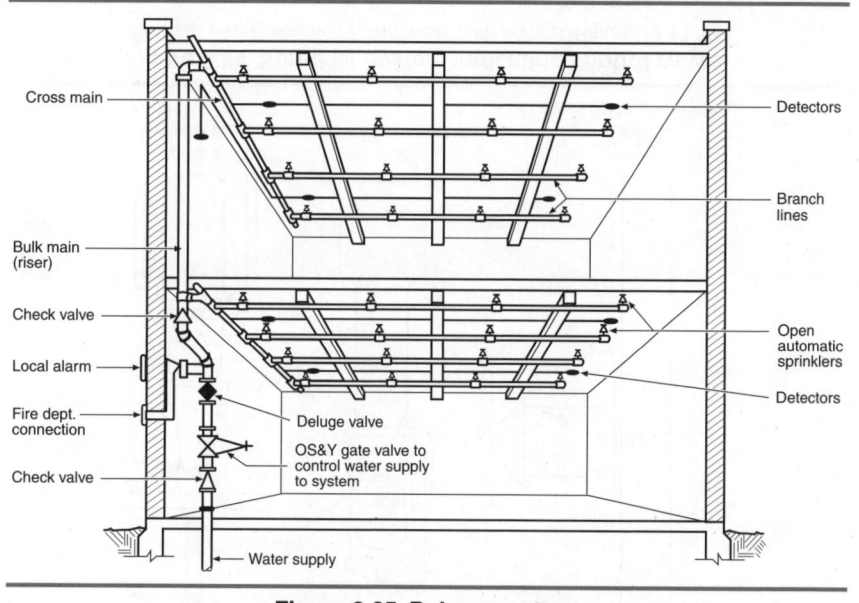

Figure 3.85 Deluge system.
Source: *Fire Protection Handbook,* 19th edition, Figure 10.11.13.

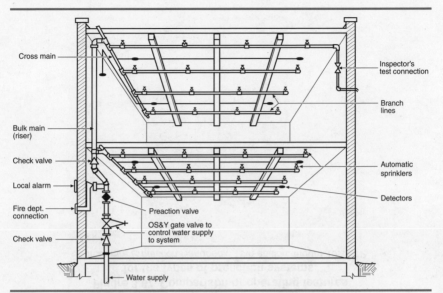

Figure 3.86 Typical preaction sprinkler system.
Source: *Fire Protection Handbook,* 19th edition, Figure 10.11.11.

BUILDING CONSTRUCTION AND SYSTEMS
Fire Protection Systems

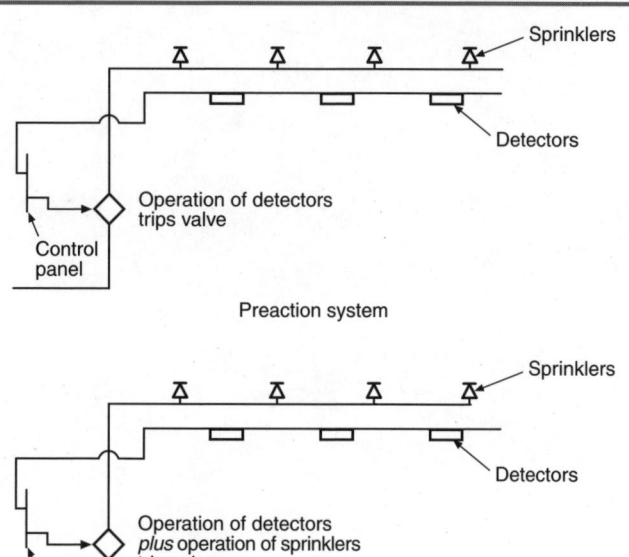

Preaction system

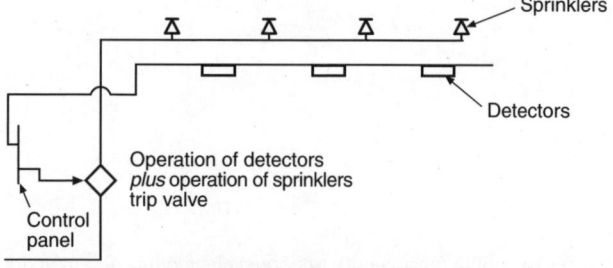

Double-interlock preaction system

**Figure 3.87 Comparison of operating features
for the types of preaction systems.**
Source: *Fire Protection Handbook,* 19th edition, Figure 10.11.12.

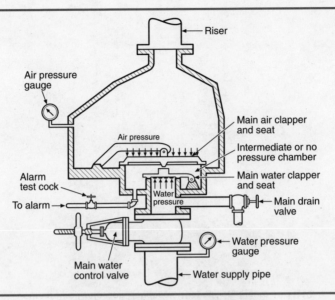

Figure 3.88 Typical dry-pipe valve.
Source: *Fire Protection Handbook,* 19th edition, Figure 10.11.9.

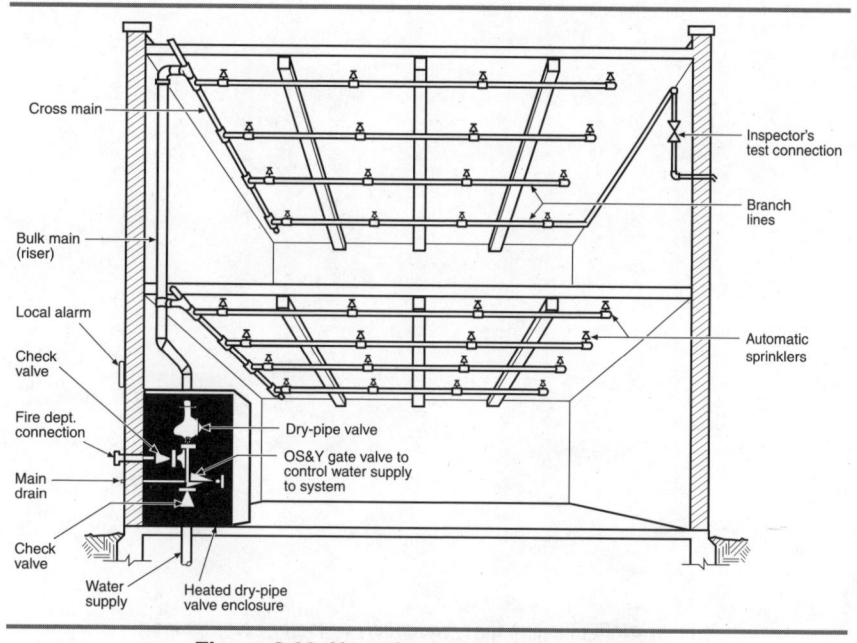

Figure 3.89 Hypothetical dry-pipe system.
Source: *Fire Protection Handbook,* 19th edition, Figure 10.11.10.

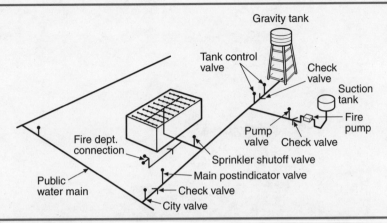

Figure 3.90 Hypothetical sprinkler system installation illustrating various water supply sources and system attachments.
Source: *Fire Protection Handbook,* 19th edition, Figure 10.11.7.

BUILDING CONSTRUCTION AND SYSTEMS
Fire Protection Systems

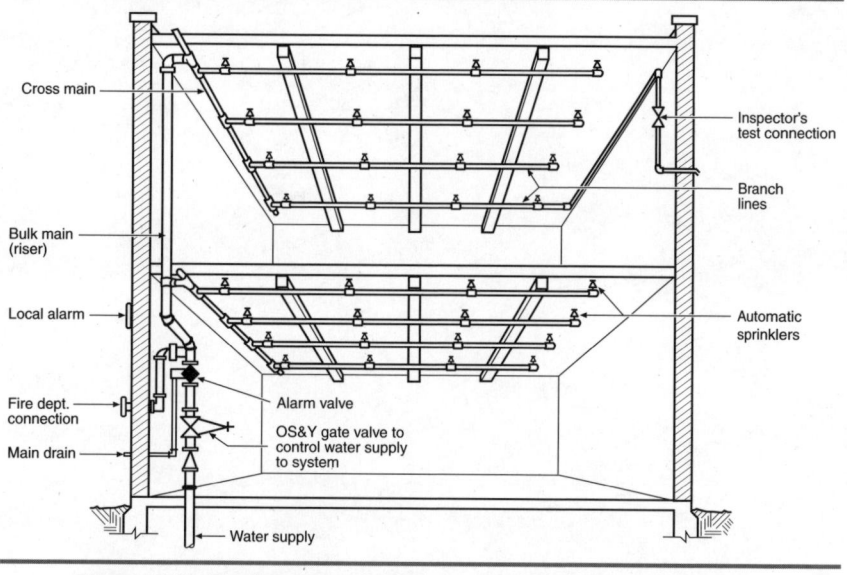

Figure 3.91 Basic components of a wet-pipe sprinkler system.
Source: *Fire Protection Handbook,* 19th edition, Figure 10.11.8.

PART 4

INFORMATION FOR THE FIRE INVESTIGATOR

Part 4 of the *Field Guide for Fire Investigators* provides data about ignition sources and materials for the fire investigator to consult either in the field or after returning from the fire scene. This Part includes the following elements:

- Properties of Ignition Sources
- Properties of Materials
- Burn Injury and Overpressure Damage
- Other Basic Information for the Fire Investigator
- Selected Metric Information for the Fire Investigator

PROPERTIES OF IGNITION SOURCES

Tables 4.1 and 4.2 present information on spontaneous heating and ignition, respectively.

Table 4.1 Materials Subject to Spontaneous Heating

(Originally prepared by the NFPA Committee on Spontaneous Heating and Ignition, which has been discontinued. Omission of any material does not necessarily indicate that it is not subject to spontaneous heating.)

Name	Tendency to Spontaneous Heating	Usual Shipping Container or Storage Method	Precautions Against Spontaneous Heating	Remarks
Alfalfa meal	High	Bags, bulk	Avoid moisture extremes. Tight cars for transportation are essential.	Many fires attributed to spontaneous heating probably caused by sparks, burning embers, or particles of hot metal picked up by the meal during processing. Test fires caused in this manner have smoldered for 72 hr before becoming noticeable.
Burlap bags "Used"	Possible	Bales	Keep cool and dry.	Tendency to heat dependent on previous use of bags. If oily would be dangerous.
Castor oil	Very slight	Metal barrels, metal cans in wooden boxes	Avoid contact of leakage from containers with rags, cotton, or other fibrous combustible materials.	Possible heating of saturated fabrics in badly ventilated piles.
Charcoal	High	Bulk, bags	Keep dry. Supply ventilation.	Hardwood charcoal must be carefully prepared and aged. Avoid wetting and subsequent drying.
Coal, bituminous	Moderate	Bulk	Store in small piles. Avoid high temperatures.	Tendency to heat depends upon origin and nature of coals. High volatile coals are particularly liable to heat.
Cocoa bean shell tankage	Moderate	Burlap bags, bulk	Observe extreme caution to maintain safe moisture limits.	This material is very hygroscopic and is liable to heating if moisture content is excessive. Precaution should be observed to maintain dry storage, etc.
Cocoanut oil	Very slight	Drums, cans, glass	Avoid contact of leakage from containers with rags, cotton, or other fibrous combustible materials.	Only dangerous if fabrics, etc. are impregnated.

Continued

Table 4.1 Materials Subject to Spontaneous Heating (continued)

Name	Tendency to Spontaneous Heating	Usual Shipping Container or Storage Method	Precautions Against Spontaneous Heating	Remarks
Cod liver oil	High	Drums, cans, glass	Avoid contact of leakage from containers with rags, cotton, or other fibrous combustible materials.	Impregnated organic materials are extremely dangerous.
Colors in oil	High	Drums, cans, glass	Avoid contact of leakage from containers with rags, cotton, or other fibrous combustible materials.	May be very dangerous if fabrics, etc., are impregnated.
Copra	Slight	Bulk	Keep cool and dry.	Heating possible if wet and hot.
Corn-meal feeds	High	Burlap bags, paper bags, bulk	Process material carefully to maintain safe moisture content and to cure before storage.	Usually contains an appreciable quantity of oil which has rather severe tendency to heat.
Corn oil	Moderate	Barrels, tank cars	Avoid contact of leakage from containers with rags, cotton, or other fibrous combustible materials.	Dangerous heating of meals, etc., unlikely unless stored in large piles while hot.
Cottonseed	Low	Bags, bulk	Keep cool and dry.	Heating possible if piled wet and hot.
Cottonseed oil	Moderate	Barrels, tank cars	Avoid contact of leakage from containers with rags, cotton, or other fibrous combustible materials.	May cause heating of saturated material in badly ventilated piles.
Distillers' dried grains with oil content (Brewers' grains)	Moderate	Bulk	Maintain moisture 7 percent to 10 percent. Cool below 100°F (38°C) before storage.	Very dangerous if moisture content is 5 percent or lower.
No oil content	Moderate	Bulk	Maintain moisture 7 percent to 10 percent. Cool below 100°F (38°C) before storage.	Very dangerous if moisture content is 5 percent or lower.
Feeds, various	Moderate	Bulk, bags	Avoid extremely low or high moisture content.	Ground feeds must be carefully processed. Avoid loading or storing unless cooled.

Continued

Table 4.1 Materials Subject to Spontaneous Heating (continued)

Name	Tendency to Spontaneous Heating	Usual Shipping Container or Storage Method	Precautions Against Spontaneous Heating	Remarks
Fertilizers				
organic, inorganic, combination of both mixed,	Moderate	Bulk, bags	Avoid extremely low or high moisture content.	Organic fertilizers containing nitrates must be carefully prepared to avoid combinations that might initiate heating.
synthetic, containing nitrates and organic matter	Moderate	Bulk, bags	Avoid free acid in preparation.	Insure ventilation in curing process by small piles or artificial drafts. If stored or loaded in bags, provide ventilation space between bags.
Fish meal	High	Bags, bulk	Keep moisture 6 percent to 12 percent. Avoid exposure to heat.	Dangerous if overdried or packaged over 100°F (38°C)
Fish oil	High	Barrels, drums, tank cars	Avoid contact of leakage from containers with rags, cotton, or other fibrous combustible materials.	Impregnated porous or fibrous materials are extremely dangerous. Tendency of various fish oils to heat varies with origin.
Fish scrap	High	Bulk, bags	Avoid moisture extremes.	Scrap loaded or stored before cooling is extremely liable to heat.
Foam rubber in consumer products	Moderate		Where possible remove foam rubber pads, etc., from garments to be dried in dryers or over heaters. If garments containing foam rubber parts have been artificially dried, they should be thoroughly cooled before being piled, bundled, or put away. Keep heating pads, hair dryers, other heat sources from contact with foam rubber pillows, etc.	Foam rubber may continue to heat spontaneously after being subjected to forced drying as in home or commercial dryers, and after contact with heating pads and other heat sources. Natural drying does not cause spontaneous heating.
Grain (various kinds)	Very slight	Bulk, bags	Avoid moisture extremes.	Ground grains may heat if wet and warm.

Continued

Table 4.1 Materials Subject to Spontaneous Heating *(continued)*

Name	Tendency to Spontaneous Heating	Usual Shipping Container or Storage Method	Precautions Against Spontaneous Heating	Remarks
Hay	Moderate	Bulk, bales	Keep dry and cool.	Wet or improperly cured hay is almost certain to heat in hot weather. Baled hay seldom heats dangerously.
Hides	Very slight	Bales	Keep dry and cool.	Bacteria in untreated hides may initiate heating.
Iron pyrites	Moderate	Bulk	Avoid large piles. Keep dry and cool.	Moisture accelerates oxidation of finely divided pyrites.
Lard oil	Slight	Wooden barrels	Avoid contact of leakage from containers with rags, cotton, or other fibrous combustible materials.	Dangerous on fibrous combustible substances.
Istle	Very slight	Bulk, bales	Keep cool and dry.	Heating possible in wet material. Unlikely under ordinary conditions. Partially burned or charred fiber is dangerous.
Jute	Very slight	Bulk	Keep cool and dry.	Avoid storing or loading in hot wet piles. Partially burned or charred material is dangerous.
Lamp black	Very slight	Wooden cases	Keep cool and dry.	Fires most likely to result from sparks or included embers, etc., rather than spontaneous heating.
Lanolin	Negligible	Glass, cans, metal drums, barrels	Avoid contact of leakage from containers with rags, cotton, or other fibrous combustible materials.	Heating possible on contaminated fibrous matter.
Lime, unslaked (calcium oxide, pebble lime, quicklime)	Moderate	Paper bags, wooden barrels, bulk	Keep dry. Avoid hot loading.	Wetted lime may heat sufficiently to ignite wood containers, etc.
Linseed	Very slight	Bulk	Keep cool and dry.	Tendency to heat dependent on moisture and oil content.

Continued

Table 4.1 Materials Subject to Spontaneous Heating (continued)

Name	Tendency to Spontaneous Heating	Usual Shipping Container or Storage Method	Precautions Against Spontaneous Heating	Remarks
Linseed oil	High	Tank cars, drums, cans, glass	Avoid contact of leakage from containers with rags, cotton, or other fibrous combustible materials.	Rags or fabrics impregnated with this oil are extremely dangerous. Avoid piles, etc. Store in closed containers, preferably metal.
Manure	Moderate	Bulk	Avoid extremes of low or high moisture contents. Ventilate the piles.	Avoid storing or loading uncooled manures.
Menhaden oil	Moderate to high	Barrels, drums, tank cars	Avoid contact of leakage from containers with rags, cotton, or other fibrous combustible materials.	Dangerous on fibrous product.
Metal powders[a]	Moderate	Drums, etc.	Keep in closed containers.	Moisture accelerates oxidation of most metal powders.
Metal turnings[a]	Practically none	Bulk	Not likely to heat spontaneously.	Avoid exposure to sparks.
Mineral wool	None	Pasteboard boxes, paper bags	Noncombustible. If loaded hot may ignite containers and other combustible surroundings.	This material is mentioned in this table only because of general impression that it heats spontaneously.
Mustard oil, black	Low	Barrels	Avoid contact of leakage with rags, cotton, or other fibrous combustible materials.	Avoid contamination of fibrous combustible materials.
Oiled clothing	High	Fiber boxes	Dry thoroughly before packing.	Dangerous if wet material is stored in piles without ventilation.
Oiled fabrics	High	Rolls	Keep ventilated. Dry thoroughly before packing.	Improperly dried fabrics extremely dangerous. Tight rolls are comparatively safe.
Oiled rags	High	Bales	Avoid storing in bulk in open.	Dangerous if wet with drying oil.
Oiled silk	High	Fiber boxes, rolls	Supply sufficient ventilation.	Improperly dried material is dangerous in form of piece goods. Rolls relatively safe.

Continued

Table 4.1 Materials Subject to Spontaneous Heating (continued)

Name	Tendency to Spontaneous Heating	Usual Shipping Container or Storage Method	Precautions Against Spontaneous Heating	Remarks
Oleic acid	Very slight	Glass bottles, wooden barrels	Avoid contact of leakage from containers with rags, cotton, or other fibrous combustible materials.	Impregnated fibrous materials may heat unless ventilated.
Oleo oil	Very slight	Wooden barrels	Avoid contact of leakage from containers with rags, cotton, or other fibrous combustible materials.	May heat on impregnated fibrous combustible matter.
Olive oil	Moderate to low	Tank cars, drums, cans, glass	Avoid contact of leakage from containers with rags, cotton, or other fibrous combustible materials.	Impregnated fibrous materials may heat unless ventilated. Tendency varies with origin of oil.
Paint containing dry-ing oil	Moderate	Drums, cans, glass	Avoid contact of leakage from containers with rags, cotton, or other fibrous combustible materials.	Fabrics, rags, etc., impregnated with paints that contain drying oils and driers are extremely danger-ous. Store in closed containers, preferably metal.
Paint scrapings	Moderate	Barrels, drums	Avoid large unventilated piles.	Tendency to heat depends on state of dryness of the scrapings.
Palm oil	Low	Wooden barrels	Avoid contact of leakage from containers with rags, cotton, or other fibrous combustible materials.	Impregnated fibrous materials may heat unless ven-tilated. Tendency varies with origin of oil.
Peanut oil	Low	Wooden barrels, tin cans	Avoid contact of leakage from containers with rags, cotton, or other fibrous combustible materials.	Impregnated fibrous materials may heat unless ven-tilated. Tendency varies with origin of oil.
Peanuts, "red skin"	High	Paper bags, cans, fiber board boxes, burlap bags	Avoid badly ventilated storage.	This is the part of peanut between outer shell and peanut itself. Provide well ventilated storage.
Peanuts, shelled	Very slight or negligible	Paper bags, cans, fiber board boxes, burlap bags	Keep cool and dry.	Avoid contamination of rags, etc., with oil.

Continued

Table 4.1 Materials Subject to Spontaneous Heating (continued)

Name	Tendency to Spontaneous Heating	Usual Shipping Container or Storage Method	Precautions Against Spontaneous Heating	Remarks
Perilla oil	Moderate to high	Tin cans, barrels	Avoid contact of leakage from containers with rags, cotton, or other fibrous combustible materials.	Impregnated fibrous materials may heat unless ventilated. Tendency varies with origin of oil.
Pine oil	Moderate	Glass, drums	Avoid contact of leakage from containers with rags, cotton, or other fibrous combustible materials.	Impregnated fibrous materials may heat unless ventilated. Tendency varies with origin of oil.
Powdered eggs	Very slight	Wooden barrels	Avoid conditions that promote bacterial growth. Inhibit against decay. Keep cool.	Possible heating of decaying powder in storage.
Powdered milk	Very slight	Wooden and fiber boxes, metal cans	Avoid conditions that promote bacterial growth. Inhibit against decay. Keep cool.	Possible heating by decay or fermentation.
Rags	Variable	Bales	Avoid contamination with drying oils. Avoid charring. Keep cool and dry.	Tendency depends on previous use of rags. Partially burned or charred rags are dangerous.
Red oil	Moderate	Glass bottles, wooden barrels	Avoid contact of leakage from containers with rags, cotton, or other fibrous combustible materials.	Impregnated porous or fibrous materials are extremely dangerous. Tendency varies with origin of oil.
Roofing felts and papers	Moderate	Rolls, bales, crates	Avoid over-drying the material. Supply ventilation.	Felts, etc., should have controlled moisture content. Packaging or rolling uncooled felts is dangerous.
Sawdust	Possible	Bulk	Avoid contact with drying oils. Avoid hot, humid storage.	Partially burned or charred sawdust may be dangerous.
Scrap film (Nitrate)	Very slight	Drums and lined boxes	Film must be properly stabilized against decomposition.	Nitrocellulose film ignites at low temperature. External ignition more likely than spontaneous heating. Avoid exposure to sparks, etc.
Scrap leather	Very slight	Bales, bulk	Avoid contamination with drying oils.	Oil-treated leather scraps may heat.

Continued

Table 4.1 Materials Subject to Spontaneous Heating (continued)

Name	Tendency to Spontaneous Heating	Usual Shipping Container or Storage Method	Precautions Against Spontaneous Heating	Remarks
Scrap rubber or buffings	Moderate	Bulk, drums	Buffings of high rubber content should be shipped and stored in tight containers.	Sheets, slabs, etc., are comparatively safe unless loaded or stored before cooling thoroughly.
Sisal	Very slight	Bulk, bales	Keep cool and dry.	Partially burned or charred material is particularly liable to ignite spontaneously.
Soybean oil	Moderate	Tin cans, barrels, tank cars	Avoid contact with rags, cotton, or fibrous materials.	Impregnated fibrous materials may heat unless well ventilated.
Sperm oil—see whale oil				
Tankage	Variable	Bulk	Avoid extremes of moisture contents. Avoid loading or storing while hot.	Very dry or moist tankages often heat. Tendency more pronounced if loaded or stored before cooling.
Tung nut meals	High	Paper bags, bulk	Material must be very carefully processed and cooled thoroughly before storage.	These meals contain residual oil which has high tendency to heat. Material also susceptible to heating if over-dried.
Tung oil	Moderate	Tin cans, barrels, tank cars	Avoid contact of leakage from containers with rags, cotton, or other fibrous combustible materials.	Impregnated fibrous materials may heat unless ventilated. Tendency varies with origin of oil.
Turpentine	Low	Tin, glass, barrels	Avoid contact of leakage from containers with rags, cotton, or other fibrous combustible materials.	Has some tendency to heat but less so than the drying oils. Chemically active with chlorine compounds and may cause fire.
Varnished fabrics	High	Boxes	Process carefully. Keep cool and ventilated.	Thoroughly dried varnished fabrics are comparatively safe.
Wallboard	Slight	Wrapped bundles, pasteboard boxes	Maintain safe moisture content. Cool thoroughly before storage.	This material is entirely safe from spontaneous heating if properly processed.

Continued

Table 4.1 Materials Subject to Spontaneous Heating (continued)

Name	Tendency to Spontaneous Heating	Usual Shipping Container or Storage Method	Precautions Against Spontaneous Heating	Remarks
Waste paper	Moderate	Bales	Keep dry and ventilated.	Wet paper occasionally heats in storage in warm locations.
Whale oil	Moderate	Barrels and tank cars	Avoid contact of leakage from containers with rags, cotton, or other fibrous combustible materials.	Impregnated fibrous materials may heat unless ventilated. Tendency varies with origin of oil.
Wool wastes	Moderate	Bulk, bales, etc.	Keep cool and ventilated or store in closed containers. Avoid high moisture.	Most wool wastes contain oil, etc., from the weaving and spinning and are liable to heat in storage. Wet wool wastes are very liable to spontaneous heating and possible ignition.

[a]Refers to iron, steel, brass, aluminum, and other common metals.
Source: *Fire Protection Handbook*, 19th edition, Table A.10.

Table 4.2 Some Materials Subject to Spontaneous Ignition

Material	Tendency	Material	Tendency
Charcoal	High	Manure	Moderate
Fish meal	High	Wool wastes	Moderate
Linseed oiled rags	High	Baled rags	Variable (low to moderate)
Brewing grains	Moderate	Sawdust	Possible
Latex foam rubber	Moderate	Grain	Low
Hay	Moderate		

Source: NFPA 921, 2001 edition, Table 3.3.5.

Table 4.3 presents the temperatures of some ignition sources. Note that although the temperature of a source may be above the ignition temperature of a material, a brief exposure of the fuel to the source may not be sufficient to result in ignition. Table 4.4 displays the ignition properties of selected materials of interest to the fire investigator.

For further discussion of ignition, see Chapter 3 of NFPA 921.

Table 4.3 Reported Burning and Sparking Temperatures of Selected Ignition Sources

Source	Temperature °F	Temperature °C	Source	Temperature °F	Temperature °C
Flames			Embers[d]		
Benzene[a]	1690	920	Cigarette (puffing)	1520–1670	830–910
Gasoline[a]	1879	1026	Cigarette (free burn)	930–1300	500–700
JP-4[b]	1700	927	Mechanical sparks[e]		
Kerosene[a]	1814	990	Steel tool	2550	1400
Methanol[a]	2190	1200	Copper–nickel alloy	570	300
Wood[c]	1880	1027			

[a]From Drysdale, *An Introduction to Fire Dynamics.*
[b]From Hagglund and Persson, *The Heat Radiation from Petroleum Fires.*
[c]From Hagglund and Persson, *An Experimental Study of the Radiation from Wood Flames.*
[d]From Krasny, *Cigarette Ignition of Soft Furnishings—A Literature Review with Commentary.*
[e]From NFPA *Fire Protection Handbook,* 15th ed., pp. 4–167.
Source: NFPA 921, 2001 edition, Table 3.3.

INFORMATION FOR THE FIRE INVESTIGATOR
Properties of Ignition Sources

Table 4.4 Ignition Properties of Selected Materials

Material	Ignition Temperature °F	Ignition Temperature °C	Minimum Radiant Flux (kW/m^2)	Energy Required (kJ/m^2)	Minimum Ignition Energy (mJ)
Solids					
Polyethylene[a]	910	488	19	1500–5100	—
Polystyrene[a]	1063	573	29	1300–6400	—
Polyurethane (flexible)[a]	852–1074	456–579	16–30	150–770	—
PVC[a]	945	507	21	3320	—
Softwood[b]	608–660	320–350	—	—	—
Hardwood[b]	595–740	313–393	—	—	—
Dusts (cloud)[c]					
Aluminum	1130	610	—	—	10
Coal	1346	730	—	—	100
Grain	805	430	—	—	30
Liquids[d]					
Acetone	869	465	—	—	1.15[e]
Benzene	928	498	—	—	0.22[e]
Ethanol	685	363	—	—	—
Gasoline (100 octane)	853	456	—	—	—
Kerosene	410	210	—	—	—
Methanol	867	464	—	—	0.14[e]
Methyl ethyl ketone	759	404	—	—	0.53[e]
Toluene	896	480	—	—	2.5[f]
Gases[d]					
Acetylene	581	305	—	—	0.02[e]
Methane	999	537	—	—	0.28[e]
Natural gas	900–1170	482–632	—	—	0.30[f]
Propane	842	450	—	—	0.25[e]

[a]From NFPA *Fire Protection Handbook,* 17th ed., Table A.6.
[b]From NFPA *Fire Protection Handbook,* 17th ed., pp. 3–25.
[c]From NFPA *Fire Protection Handbook,* 16th ed., Table 5.9A.
[d]Ignition temperatures from *NFPA Fire Protection Guide to Hazardous Materials.*
[e]From the *SFPE Handbook of Fire Protection Engineering,* Table 2-5.2.
[f]From NFPA *Fire Protection Handbook,* 15th ed., Table 11.3B.
Source: NFPA 921, 2001 edition, Table 3.3.4.

PROPERTIES OF MATERIALS

This section contains a selection of relevant data and properties of solid liquid and gaseous materials relating to ignition and flammability. Subjects covered include ignition temperatures, flammable limits, rates of heat release, heats of combustion, melting temperatures, and many others. Since there are many data tables, it may be necessary to refer to more than one to find all the information for a given material or item.

In some instances, values may vary somewhat from source to source. This may be due to differences in the materials tested or the method of test used. The values should be considered as general guidance and if exact values are needed, the material or materials in question should be tested.

Properties of Solids

Tables 4.5 through 4.14 present selected properties of solids. Note that Table 4.7 considers plastics as solids.

Table 4.5 Thermal Properties of Selected Materials

Material	Thermal Conductivity (k) (W/m-K)	Density (ρ) (kg/m^3)	Heat Capacity (c_p) (J/kg-K)
Copper	387	8940	380
Concrete	0.8–1.4	1900–2300	880
Gypsum plaster	0.48	1440	840
Oak	0.17	800	2380
Pine (yellow)	0.14	640	2850
Polyethylene	0.35	940	1900
Polystyrene (rigid)	0.11	1100	1200
Polyvinylchloride	0.16	1400	1050
Polyurethane[a]	0.034	20	1400

[a]Typical values, properties vary.

Source: NFPA 921, 2001 edition, Table 3.2.1.

Table 4.6 Heats of Combustion and Related Properties of Pure Substances

Material	Composition	Molecular Weight, W	Gross, Δh_c^u (MJ/kg)	Net, Δh_c (MJ/kg)	$\Delta h_c'/r_o$ (MJ/kg O_2)	Oxygen Fuel Mass Ratio, r_o	Boiling Temp., T_b (°C)	Latent Heat of Vaporization, Δh_v (kJ/kg)	Liquid Heat Capacity, C_{pl} (kJ/kg·°C)	Vapor Heat Capacity, C_{pv} (kJ/kg·°C)
Acetaldehyde	C_2H_4O	44.05	27.07	25.07	13.81	1.816	20.8	—	1.94	1.24
Acetic acid	$C_2H_4O_2$	60.05	14.56	13.09	12.28	1.066	118.1	395	—	1.11
Acetone	C_3H_6O	58.08	30.83	28.56	12.96	2.204	56.5	501	2.12	1.29
Acetylene	C_2H_2	26.04	49.91	48.22	15.70	3.072	-84.0	—	—	1.69
Acrolein	C_3H_4O	56.06	29.08	27.51	13.77	1.998	52.5	505	—	1.17
Acrylonitrile	C_3H_3N	53.06	33.16	31.92	14.11	2.262	77.3	615	2.10	1.20
(Allene) → propadiene										
Ammonium perchlorate[a]	NH_4ClO_4	117.49	2.35	2.16	3.97	0.545	—	—	—	—
iso-Amyl alcohol	$C_5H_{12}O$	88.15	37.48	34.49	12.67	2.723	132.0	501	2.90	1.50
Aniline	C_6H_7N	93.12	36.44	34.79	13.06	2.663	184.4	478	2.08	1.16
Benzaldehyde	C_7H_6O	106.12	33.25	32.01	13.27	2.412	179.2	385	1.61	—
Benzene	C_6H_6	78.11	41.83	40.14	13.06	3.073	80.1	389	1.72	1.05
Benzoic acid[a]	$C_7H_6O_2$	122.12	26.43	25.35	12.90	1.965	250.8	415	—	0.85
Benzyl alcohol	C_7H_8O	108.13	34.56	32.93	13.09	2.515	205.7	467	2.00	1.19
Bicyclohexyl	$C_{12}H_{22}$	166.30	45.35	42.44	12.61	3.367	236.0	263	—	—
1,2-Butadiene	C_4H_6	54.09	47.95	45.51	13.99	3.254	10.8	—	—	1.48
1,3-Butadiene	C_4H_6	54.09	46.99	44.55	13.69	3.254	-4.4	—	—	1.47
(1,3-Butadiyne) → diacetylene										
n-Butane	C_4H_{10}	58.12	49.50	45.72	12.77	3.579	-0.5	—	2.30	1.68
iso-Butane	C_4H_{10}	58.12	48.95	45.17	12.62	3.579	-11.8	—	—	1.67
1-Butene	C_4H_8	56.10	48.44	45.31	13.24	3.422	-6.2	—	—	1.53
n-Butylamine	$C_4H_{11}N$	73.14	41.75	38.45	12.84	2.994	77.8	372	2.57	1.62
d-Camphor[a]	$C_{10}H_{16}O$	152.23	38.75	36.44	12.84	2.838	203.4	—	—	0.82
Carbon[a]	C	12.01	32.80	32.80	12.31	2.664	4200.0	—	—	0.71

Continued

Table 4.6 Heats of Combustion and Related Properties of Pure Substances (continued)

Material	Composition	Molecular Weight, W	Gross, Δh_c^u (MJ/kg)	Net, $\Delta h_c'$ (MJ/kg)	$\Delta h_c'/r_o$ (MJ/kg O_2)	Oxygen Fuel Mass Ratio, r_o	Boiling Temp., T_b (°C)	Latent Heat of Vaporization, Δh_v (kJ/kg)	Liquid Heat Capacity, c_{pl} (kJ/kg·°C)	Vapor Heat Capacity, c_{pv} (kJ/kg·°C)
Carbon disulfide	CS_2	76.13	6.34	6.34	5.03	1.261	46.5	351	1.00	0.60
Carbon monoxide	CO	28.01	10.10	10.10	17.69	0.571	−191.3	—	—	1.04
Cellulose[a]	$C_6H_{10}O_5$	162.14	17.47	16.12	13.61	1.184	—	—	1.16	—
(Chloroethylene) → vinyl chloride										
(Chloroform) → trichloromethane										
Chlorotrifluoroethylene	C_2F_3Cl	116.47	2.00	2.00	3.64	0.549	−28.3	188	1.34	0.72
m-Cresol	C_7H_8O	108.13	34.26	32.64	12.98	2.515	202.2	399	2.00	1.13
Cumene	C_9H_{12}	120.19	43.40	41.20	12.90	3.195	152.3	312	1.77	1.26
Cyanogen	C_2N_2	52.04	21.06	21.06	17.12	1.230	−21.2	—	—	1.12
Cyclobutane	C_4H_8	56.10	48.91	45.77	13.38	3.422	12.9	—	—	1.29
Cyclohexane	C_6H_{12}	84.16	46.58	43.45	12.70	3.422	80.7	357	1.84	1.26
Cyclohexene	C_6H_{10}	82.14	45.67	42.99	12.99	3.311	82.8	371	1.80	1.28
Cyclohexylamine	$C_6H_{13}N$	99.18	41.05	38.17	12.79	2.984	134.5	—	—	—
Cyclopentane	C_5H_{10}	70.13	46.93	43.80	12.80	3.422	49.3	389	2.23	1.18
Cyclopropane	C_3H_6	42.08	49.70	46.57	13.61	3.422	−32.9	—	1.92	1.33
(Decahydronaphthalene) → cis-decalin										
cis-Decalin	$C_{10}H_{18}$	138.24	45.49	42.63	12.70	3.356	195.8	309	1.67	1.21
n-Decane	$C_{10}H_{22}$	142.28	47.64	44.24	12.69	3.486	174.1	276	2.19	1.85
Diacetylene	C_4H_2	50.06	46.60	45.72	15.89	2.877	10.3	—	—	1.47
(Diamine) → hydrazine										
Diborane	H_6B_2	27.69	79.80	79.80	23.02	3.467	−92.5	—	—	1.75
Dichloromethane	CH_2Cl_2	84.94	6.54	6.02	10.65	0.565	39.7	330	1.18	0.80
Diethyl cyclohexane	$C_{10}H_{20}$	140.26	46.30	43.17	12.58	3.422	174.0	—	1.87	—
Diethyl ether	$C_4H_{10}O$	74.12	36.75	33.79	13.04	2.590	34.6	360	2.34	1.52

Continued

Table 4.6 Heats of Combustion and Related Properties of Pure Substances (continued)

Material	Composition	Molecular Weight, W	Gross, Δh^u_c (MJ/kg)	Net, Δh^i_c (MJ/kg)	$\Delta h^i_c / r_o$ (MJ/kg O_2)	Oxygen Fuel Mass Ratio, r_o	Boiling Temp., T_b (°C)	Latent Heat of Vaporization, Δh_v (kJ/kg)	Liquid Heat Capacity, C_{pl} (kJ/kg·°C)	Vapor Heat Capacity, C_{pv} (kJ/kg·°C)
(2,4 Diisocyanotoulene) → toluene diisocyanate										
(Diisopropyl ether) → iso-propyl ether										
Dimethylamine	C_2H_7N	45.08	38.66	35.25	13.24	2.662	6.9	—	—	1.80
(Dimethyl aniline) → xylidene										
Dimethyldecalin	$C_{12}H_{22}$	166.30	45.70	42.79	13.15	3.254	220.0	260	—	
(Dimethyl ether) → methyl ether										
1,1-Dimethylhydrazine (UDMH)	$C_2H_8N_2$	60.10	32.95	30.03	14.10	2.130	25.0	578	2.73	
Dimethyl sulfoxide	C_2H_6SO	78.13	29.88	28.19	15.30	1.843	189.0	677	1.89	1.14
1,3 Dioxane	$C_4H_8O_2$	88.10	26.57	24.58	9.66	2.543	105.0	404	—	
1,4 Dioxane	$C_4H_8O_2$	88.10	26.83	24.84	9.77	2.543	101.1	406	1.74	1.07
Ethane	C_2H_6	30.07	51.87	47.49	12.75	3.725	−88.6	—	—	1.75
Ethanol	C_2H_6O	46.07	29.67	26.81	12.87	2.084	78.5	837	2.43	1.42
(Ethene) → ethylene										
Ethyl acetate	$C_4H_8O_2$	88.10	25.41	23.41	12.89	1.816	77.2	367	1.94	1.29
Ethyl acrylate	$C_5H_8O_2$	100.12	27.44	25.69	13.39	1.918	100.0	290	—	1.14
Ethylamine	C_2H_7N	45.08	38.63	35.22	13.23	2.662	16.5	—	2.89	1.61
Ethyl benzene	C_8H_{10}	106.16	43.00	40.93	12.93	3.165	136.1	339	1.75	1.21
Ethylene	C_2H_4	28.05	50.30	47.17	13.78	3.422	−103.9	—	2.38	1.56
Ethylene glycol	$C_2H_6O_2$	62.07	19.17	17.05	13.22	1.289	197.5	800	2.43	1.56
Ethylene oxide	C_2H_4O	44.05	29.65	27.65	15.23	1.816	10.7	—	1.97	1.10
(Ethylene trichloride) → trichloroethylene										
(Ethyl ether) → diethyl ether										
Formaldehyde	CH_2O	30.03	18.76	17.30	16.23	1.066	−19.3	—	—	1.18
Formic acid	CH_2O_2	46.03	5.53	4.58	13.15	0.348	100.5	476	2.15	0.98
Furan	C_4H_4O	68.07	30.61	29.32	13.86	2.115	31.4	398	1.69	0.96

Continued

Table 4.6 Heats of Combustion and Related Properties of Pure Substances (continued)

Material	Composition	Molecular Weight, W	Gross, Δh_c^o (MJ/kg)	Net, Δh_c^o (MJ/kg)	$\Delta h_c^o/r_o$ (MJ/kg O_2)	Oxygen Fuel Mass Ratio, r_o	Boiling Temp., T_b (°C)	Latent Heat of Vaporization, Δh_v (kJ/kg)	Liquid Heat Capacity, C_{pl} (kJ/kg·°C)	Vapor Heat Capacity, C_{pv} (kJ/kg·°C)
a-D-glucose[a]	$C_6H_{12}O_6$	180.16	15.55	14.08	13.21	1.066	—	—	—	—
(Glycerine) → glycerol										
Glycerol	$C_3H_8O_3$	92.10	17.95	16.04	13.19	1.216	290.0	800	2.42	1.25
(Glycerol trinitrate) → nitroglycerin										
n-Heptane	C_7H_{16}	100.20	48.07	44.56	12.68	3.513	98.4	316	2.20	1.66
n-Heptene	C_7H_{14}	98.18	47.44	44.31	12.95	3.422	93.6	317	2.17	1.58
Hexadecane	$C_{16}H_{34}$	226.43	47.25	43.95	12.70	3.462	286.7	226	2.22	1.64
Hexamethyldisiloxane	$C_6H_{18}Si_2O$	162.38	38.30	35.80	15.16	2.364	100.1	192	2.01	—
(Hexamethylenetetramine) → methenamine										
n-Hexane	C_6H_{14}	86.17	48.31	44.74	12.68	3.528	68.7	335	2.24	1.66
n-Hexene	C_6H_{12}	84.16	47.57	44.44	12.99	3.422	63.5	333	2.18	1.57
Hydrazine	H_4N_2	32.05	52.08	49.34	49.40	0.998	113.5	1180	3.08	1.65
Hydrazoic acid	HN_3	43.02	15.28	14.77	79.40	0.186	35.7	690	—	1.02
Hydrogen	H_2	2.00	141.79	130.80	16.35	8.000	-252.7	—	—	14.42
(Hydrogen azide) → hydrazoic acid										
Hydrogen cyanide	HCN	27.03	13.86	13.05	8.82	1.480	25.7	933	2.61	1.33
Hydrogen sulfide	H_2S	34.08	48.54	47.25	16.77	2.817	-60.3	548	—	1.00
Maleic anhydride[a]	$C_4H_2O_3$	74.04	18.77	18.17	14.01	1.297	202.0	—	—	—
Melamine[a]	$C_3H_6N_6$	126.13	15.58	14.54	12.73	1.142	—	—	—	—
Methane	CH_4	16.04	55.50	50.03	12.51	4.000	-161.5	—	—	2.23
Methanol	CH_4O	32.04	22.68	19.94	13.29	1.500	64.8	1101	2.37	1.37
Methenamine[a]	$C_6H_{12}N_4$	140.19	29.97	28.08	13.67	2.054	—	—	—	—
2-Methoxyethanol	$C_3H_8O_2$	76.09	24.23	21.92	13.03	1.682	124.4	583	2.23	—
Methylamine	CH_5N	31.06	34.16	30.62	13.21	2.318	-6.3	—	—	1.61

Continued

Table 4.6 Heats of Combustion and Related Properties of Pure Substances (continued)

Material	Composition	Molecular Weight, W	Gross, Δh_c^o (MJ/kg)	Net, Δh_c^i (MJ/kg)	$\Delta h_c^i/r_o$ (MJ/kg O_2)	Oxygen Fuel Mass Ratio, r_o	Boiling Temp., T_b (°C)	Latent Heat of Vaporization, Δh_v (kJ/kg)	Liquid Heat Capacity, C_{pl} (kJ/kg·°C)	Vapor Heat Capacity, C_{pv} (kJ/kg·°C)
(2-Methyl 1-butanol) → iso-amyl alcohol										
(Methyl chloride) → dichloromethane										
Methyl ether	C_2H_6O	46.07	31.70	28.84	13.84	2.084	−24.9	—	—	1.43
Methyl ethyl ketone	C_4H_8O	72.10	33.90	31.46	12.89	2.441	79.6	434	2.30	1.43
1-Methylnaphthalene	$C_{11}H_{10}$	142.19	40.88	39.33	12.95	3.038	244.7	323	1.58	1.12
Methyl methacrylate	$C_5H_8O_2$	100.11	27.37	25.61	12.33	2.078	101.0	360	1.91	—
Methyl nitrate	CH_3NO_3	77.04	8.67	7.81	75.10	0.104	64.6	409	2.04	0.99
(2-Methyl propane) → iso-butane										
Naphthalene[a]	$C_{10}H_8$	128.16	40.21	38.84	12.96	2.996	217.9	—	1.18	1.03
Nitrobenzene	$C_6H_5NO_2$	123.11	25.11	24.22	14.90	1.625	210.7	330	1.52	—
Nitroglycerin	$C_3H_5N_3O_9$	227.09	6.82	6.34	—	—	Unstable	462	1.49	—
Nitromethane	CH_3NO_2	61.04	11.62	10.54	15.08	0.699	101.1	567	1.74	0.94
n-Nonane	C_9H_{20}	128.25	47.76	44.33	12.69	3.493	150.6	295	2.10	1.65
Octamethyl-cyclotetrasiloxane	$C_8H_{24}Si_4O_4$	296.62	26.90	25.10	14.56	1.725	175.0	127	1.88	—
n-Octane	C_8H_{18}	114.22	47.90	44.44	12.69	3.502	125.6	301	2.20	1.65
iso-Octane	C_8H_{18}	114.22	47.77	44.31	12.65	3.502	117.7	272	2.15	1.65
1-Octene	C_8H_{10}	112.21	47.33	44.20	12.92	3.422	121.3	301	2.19	1.59
(1-Octylene) → 1-octene										
1,2-Pentadiene	C_5H_8	68.11	47.31	44.71	13.60	3.288	44.9	405	2.21	1.55
n-Pentane	C_5H_{12}	72.15	48.64	44.98	12.68	3.548	36.0	357	2.33	1.67
1-Pentene	C_5H_{10}	70.13	47.77	44.64	13.04	3.422	30.0	359	2.16	1.56
Phenol[a]	C_6H_6O	94.11	32.45	31.05	13.05	2.380	181.8	433	1.43	1.10
Phosgene	$COCl_2$	98.92	1.74	1.74	10.74	0.162	8.3	247	1.02	0.58
Propadiene	C_3H_4	40.06	48.54	46.35	14.51	3.195	−34.6	—	—	1.44
Propane	C_3H_8	44.09	50.35	46.36	12.78	3.629	−42.2	—	2.23	1.67

Continued

Table 4.6 Heats of Combustion and Related Properties of Pure Substances (continued)

Material	Molecular Composition	Molecular Weight, W	Gross, Δh_c^u (MJ/kg)	Net, Δh_c^l (MJ/kg)	$\Delta h_c^l/r_o$ (MJ/kg O$_2$)	Oxygen Fuel Mass Ratio, r_o	Boiling Temp., T_b (°C)	Latent Heat of Vaporization, Δh_v (kJ/kg)	Liquid Heat Capacity, C_{pl} (kJ/kg·°C)	Vapor Heat Capacity, C_{pv} (kJ/kg·°C)
n-Propanol	C$_3$H$_8$O	60.09	33.61	30.68	12.81	2.396	97.2	686	2.50	1.45
iso-Propanol	C$_3$H$_8$O	60.09	33.38	30.45	12.71	2.396	80.3	663	2.42	1.48
Propene	C$_3$H$_6$	42.08	48.92	45.79	13.38	3.422	-47.7	—	—	1.52
(iso-Propylbenzene) → cumene										
(Propylene) → propene										
iso-Propyl ether	C$_6$H$_{14}$O	102.17	39.26	36.25	12.86	2.819	67.8	286	2.14	1.55
Propyne	C$_3$H$_4$	40.06	48.36	46.17	14.45	3.195	-23.3	—	—	1.51
Styrene	C$_8$H$_8$	104.14	42.21	40.52	13.19	3.073	145.2	356	1.76	1.17
Sucrose[a]	C$_{12}$H$_{22}$O$_{11}$	342.30	16.49	15.08	13.44	1.122	—	—	1.24	—
(1,2,3,4-Tetrahydronaphthalene) → tetralin										
Tetralin	C$_{10}$H$_{12}$	132.20	42.60	40.60	12.90	3.147	207.0	425	1.64	1.19
Tetranitromethane	CN$_4$O$_8$	196.04	2.20	2.20	—	—	125.7	196	—	—
Toluene	C$_7$H$_8$	92.13	42.43	40.52	12.97	3.126	110.4	360	1.67	1.12
Toluene diisocyanate	C$_9$H$_6$N$_2$O$_2$	174.16	24.32	23.56	13.50	1.746	120.0	—	1.65	—
Triethanolamine	C$_6$H$_{15}$NO$_3$	149.19	29.29	27.08	15.30	1.770	360.0	—	—	—
Triethylamine	C$_6$H$_{15}$N	101.19	43.19	39.93	12.95	3.083	89.5	303	2.22	1.59
1,1,2-Trichloroethane	C$_2$H$_3$Cl$_3$	133.42	7.77	7.28	11.02	0.660	114.0	260	1.11	0.67
Trichloroethylene	C$_2$HCl$_3$	131.40	6.77	6.60	12.05	0.548	86.9	245	1.07	0.61
Trichloromethane	CHCl$_3$	119.39	3.39	3.21	9.60	0.335	61.7	249	0.97	0.55
Trinitromethane	CHN$_3$O$_6$	151.04	3.41	3.25	—	—	Unstable	—	—	—
Trinitrotoluene[a]	C$_7$H$_5$N$_3$O$_6$	227.13	15.12	14.64	19.80	0.740	240.0	322	1.40	—
Trioxane	C$_3$H$_6$O$_3$	90.08	16.57	15.11	14.17	1.066	114.5	450	—	—
Urea[a]	CH$_4$ON$_2$	60.06	10.52	9.06	11.34	0.799	—	—	—	—
Vinyl acetate	C$_4$H$_6$O$_2$	86.09	24.18	22.65	13.54	1.673	72.5	167	2.00	1.55
Vinyl acetylene	C$_4$H$_4$	52.07	47.05	45.36	14.76	3.073	5.1	—	—	1.41

Continued

Table 4.6 Heats of Combustion and Related Properties of Pure Substances (continued)

Material	Composition	Molecular Weight, W	Gross, Δh_c^u (MJ/kg)	Net, Δh_c^l (MJ/kg)	$\Delta h_c^l / r_o$ (MJ/kg O_2)	Oxygen Fuel Mass Ratio, r_o	Boiling Temp., T_b (°C)	Latent Heat of Vaporization, Δh_v (kJ/kg)	Liquid Heat Capacity, C_{pl} (kJ/kg·°C)	Vapor Heat Capacity, C_{pv} (kJ/kg·°C)
Vinyl bromide	C_2H_3Br	106.96	12.10	11.48	13.95	0.823	15.6	—	2.42	0.53
Vinyl chloride	C_2H_3Cl	62.50	20.02	16.86	11.97	1.408	−13.8	—	—	0.86
(Vinyl trichloride) → 1,1,2-trichlorolthane										
Xylenes	C_8H_{10}	106.16	42.89	40.82	12.90	3.165	138–144	343	1.72	1.21
Xylidene	$C_8H_{11}N$	121.22	38.28	36.29	12.79	2.838	192.7	366	1.77	—

[a]Denotes substance in crystalline solid form; otherwise, liquid if T_b > 25°C, gaseous if T_b > 25°C.

Source: Table C.2, *SFPE Handbook of Fire Protection Engineering*, 3rd edition, Courtesy of the Society of Fire Protection Engineers, Bethesda, MD.

Table 4.7 Heats of Combustion and Related Properties of Plastics

Material	Unit Composition	Molecular Weight, W	Gross, Δh_c^u (MJ/kg)	Net, Δh_c^j (MJ/kg)	$\Delta h_c^j/r_o$ (MJ/kg O_2)	Oxygen Fuel Mass Ratio, r_o	Heat Capacity Solid, C_{ps} (kJ/kg·°C)
Acrylonitrile-butadiene styrene copolymer	—	—	35.25	33.75			1.41–1.59
Bisphenol A epoxy	$C_{11.85}H_{20.37}O_{2.83}N_{0.3}$	212.10	33.53	31.42	13.41	2.343	
Butadiene-acrylonitrile 37% copolymer	—	—	39.94				
Butadiene/styrene 8.58% copolymer	$C_{4.18}H_{6.09}$	56.30	44.84	42.49	13.11	3.241	1.94
Butadiene/styrene 25.5% copolymer	$C_{4.60}H_{6.29}$	61.55	44.19	41.95	13.07	3.209	1.82
Cellulose acetate (triacetate)	$C_{12}H_{16}O_8$	288.14	18.88	17.66	13.25	1.333	1.34
Cellulose acetate-butyrate	$C_{12}H_{18}O_7$	274.27	23.70	22.30	14.67	1.517	1.70
Epoxy, unhardened	$C_{31}H_{36}O_{5.5}$	496.63	32.92	31.32	13.05	2.400	
Epoxy, hardened	$C_{39}H_{40}O_{8.5}$	644.74	30.27	28.90	13.01	2.221	
Melamine formaldehyde (Formica™)	$C_6H_6N_6$	162.08	19.33	18.52	12.51	1.481	1.46
Nylon 6	$C_6H_{11}NO$	113.08	30.1–31.7	28.0–29.6	12.30	2.335	1.52
Nylon 6,6	$C_{12}H_{22}N_2O_2$	226.16	31.6–31.7	29.5–29.6	12.30	2.405	1.70
Nylon 11 (Rilsan)	$C_{11}H_{21}NO$	183.14	36.99	34.47	12.33	2.796	
Phenol formaldehyde	$C_{15}H_{12}O_2$	224.17	27.9–31.6	26.7–30.4	11.80	2.427	1.70–2.30
foam			21.6–27.4	20.2–26.2			1.70
Polyacenaphthalene	$C_{12}H_8$	152.14	39.23	38.14	12.95	2.945	
Polyacrylonitrile	C_3H_3N	53.04	32.22	30.98	13.70	2.262	1.50
Polyallylphthalate	$C_{14}H_{14}O$	198.17	27.74	26.19	9.54	2.745	
(Polyamides) → nylon		—					
Poly-1,4-butadiene	C_4H_6	54.05	45.19	42.75	13.13	3.256	
Poly-1-butene	C_4H_8	56.05	46.48	43.35	12.65	3.426	1.88
Polycarbonate	$C_{16}H_{14}O_3$	254.19	30.99	29.78	13.14	2.266	1.26

Continued

Table 4.7 Heats of Combustion and Related Properties of Plastics (continued)

Material	Unit Composition	Molecular Weight, W	Gross, Δh_c^u (MJ/kg)	Net, Δh_c^l (MJ/kg)	$\Delta h_c^l / r_o$ (MJ/kg O_2)	Oxygen Fuel Mass Ratio, r_o	Heat Capacity Solid, C_{ps}^o (kJ/kg·°C)
Polycarbon suboxide	C_3O_2	68.03	13.78	13.78	14.64	0.941	
Polychlorotrifluorethylene	C_2F_3Cl	116.47	1.12	1.12	2.04	0.549	0.92
Polydiphenylbutadiene	$C_{16}H_{10}$	202.18	39.30	38.20	13.05	2.928	
Polyester, unsaturated	$C_{5.77}H_{6.25}O_{1.63}$	101.60	21.6–29.8	20.3–28.5	11.90	2.053	1.20–2.30
Polyether, chlorinated	$C_5H_8OCl_2$	154.97	17.84	16.71	12.45	1.342	
Polyethylene	C_2H_4	28.03	46.2–46.5	43.1–43.4	12.63	3.425	1.83–2.30
Polyethylene oxide	C_2H_4O	44.02	26.65	24.66	13.57	1.817	
Polyethylene terephthalate	$C_{10}H_8O_4$	192.11	22.18	21.27	12.77	1.666	1.00
Polyformaldehyde	CH_2O	30.01	16.93	15.86	14.88	1.066	1.46
Poly-1-hexene sulfone	$C_6H_{12}SO_2$	148.13	29.78	28.00	14.40	1.944	
Polyhydrocyanic acid	HCN	27.02	23.26	22.45	15.17	1.480	
(Polyisobutylene) → poly-1-butene	—						
Polyisocyanurate foam			26.30	22.2–26.2			
Polyisoprene	C_5H_8	68.06	44.90	42.30	12.90	3.291	
Poly-3-methyl-1-butene	C_5H_{10}	70.06	46.55	43.42	12.67	3.426	
Polymethyl methacrylate	$C_5H_8O_2$	100.06	26.64	24.88	12.97	1.919	1.44
Poly-4-methyl-1-pentene	C_6H_{12}	84.08	46.52	43.39	12.67	3.425	2.18
Poly-α-methylstyrene	C_9H_{10}	118.11	42.31	40.45	13.00	3.116	
Polynitroethylene	$C_2H_3O_2N$	73.03	15.96	15.06	19.64	0.767	
Polyoxymethylene	CH_2O	30.01	16.93	15.65	14.68	1.066	
Polyoxytrimethylene	C_3H_6O	58.04	31.52	29.25	13.27	2.205	
Poly-1-pentene	C_5H_{10}	70.06	45.58	42.45	12.39	3.426	

Continued

Table 4.7 Heats of Combustion and Related Properties of Plastics (continued)

Material	Unit Composition	Molecular Weight, W	Gross, Δh_c^u (MJ/kg)	Net, Δh_c^l (MJ/kg)	$\Delta h_c^l/r_o$ (MJ/kg O$_2$)	Oxygen Fuel Mass Ratio, r_o	Heat Capacity Solid, C_{ps} (kJ/kg·°C)
Polyphenylacetylene	C$_8$H$_6$	102.09	40.00	38.70	13.00	2.978	
Polyphenylene oxide	C$_8$H$_8$O	120.09	34.59	33.13	13.09	2.531	1.34
Polypropene sulfone	C$_3$H$_6$SO$_2$	106.10	23.82	22.58	16.64	1.357	
Poly-β-propiolactone	C$_3$H$_4$O$_2$	72.14	19.35	18.13	13.62	1.331	
Polypropylene	C$_3$H$_6$	42.04	46.37	43.23	12.62	3.824	2.10
Polypropylene oxide	C$_3$H$_6$O	58.04	31.17	28.90	13.11	2.205	
Polystyrene	C$_8$H$_8$	104.10	41.4-42.5	39.7-39.8	12.93	3.074	1.40
Polystyrene-foam	—		41.2-42.9	39.70			
Polystyrene-foam, FR	—		39.70	35.6-40.8			1.30
Polysulfones, butene	C$_4$H$_8$SO$_2$	120.11	24.04-26.47	22.25-25.01	14.79	1.598	
Polysulfur	S	32.06	9.72	9.72	9.74	0.998	1.02
Polytetrafluoroethylene	C$_2$F$_4$	100.02	5.00	5.00	7.81	0.640	
Polytetrahydrofuran	C$_4$H$_8$O	72.05	34.39	31.85	13.04	2.443	
Polyurea	C$_{15}$H$_{18}$O$_4$N$_4$	318.20	24.91	23.67	13.45	1.760	
Polyurethane	C$_{6.3}$H$_{7.1}$NO$_{2.1}$	130.30	23.90	22.70	13.16	1.725	1.75-1.84
Polyurethane-foam	—		26.1-31.6	23.2-28.0			
Polyurethane-foam, FR	—		24.0-25.0				
Polyvinyl acetate	C$_4$H$_6$O$_2$	86.05	23.04	21.51	12.86	1.673	
Polyvinyl alcohol	C$_2$H$_4$	44.03	25.00	23.01	12.66	1.817	1.70
Polyvinyl butyral	C$_8$H$_{14}$O$_2$	142.10	32.90	30.70	13.00	2.365	
Polyvinyl chloride	C$_2$H$_3$Cl	62.48	17.95	16.90	12.00	1.408	0.90-1.20

Continued

Table 4.7 Heats of Combustion and Related Properties of Plastics (continued)

Material	Unit Composition	Molecular Weight, W	Gross, Δh_c^u (MJ/kg)	Net, Δh_c^l (MJ/kg)	$\Delta h_c^l / r_o$ (MJ/kg O_2)	Oxygen Fuel Mass Ratio, r_o	Heat Capacity Solid, C_{ps} (kJ/kg°C)
Polyvinyl-foam	—		22.83				1.30–2.10
Polyvinyl fluoride	C_2H_3F	46.02	21.70	20.27	10.60	1.912	
Polyvinylidene chloride	$C_2H_2Cl_2$	96.93	10.52	10.07	12.21	0.825	1.34
Polyvinylidene fluoride	$C_2H_2F_2$	64.02	14.77	14.08	11.26	1.250	1.38
Urea formaldehyde	$C_3H_6O_2N_2$	102.05	15.90	14.61	13.31	1.098	1.60–2.10
Urea formaldehyde-foam	—		14.80				

Source: Table C.3, SFPE Handbook of Fire Protection Engineering, 3rd edition.
Courtesy of the Society of Fire Protection Engineers, Bethesda, MD.

INFORMATION FOR THE FIRE INVESTIGATOR
Properties of Materials

4-24

Table 4.8   Heats of Combustion of Miscellaneous Materials

Materials	Gross, Δh_c^u (MJ/kg)	Net, Δh_c^l (MJ/kg)
Acetate (see cellulose acetate)		
Acrylic fiber	30.6–30.8	
Blasting powder	2.1–2.4	
Butter	38.5	
Celluloid (cellulose nitrate and camphor)	17.5–20.6	16.4–19.2
Cellulose acetate fiber, $C_8H_{12}O_6$	17.8–18.4	16.4–17.0
Cellulose diacetate fiber, $C_{10}H_{14}O_7$	18.7	
Cellulose nitrate, $C_6H_9N_1O_7/C_6H_8N_2O_9/C_6H_7N_3O_{11}$	9.11–13.48	
Cellulose triacetate fiber, $C_{12}H_{16}O_8$	18.8	17.6
Charcoal	33.7–34.7	33.2–34.2
Coal—anthracite	30.9–34.6	30.5–34.2
—bituminous	24.7–36.3	23.6–35.2
Coke	28.0–31.0	28.0–31.0
Cork	26.1	
Cotton	16.5–20.4	
Dynamite	5.4	
Epoxy, $C_{11.9}H_{20.4}O_{2.8}N_{0.3}/C_{6.064}H_{7.550}O_{1.222}$	32.8–33.5	31.1–31.4
Fat, animal	39.8	
Flint powder	3.0–3.1	
Fuel oil—No. 1	46.1	
—No. 6	42.5	
Gasketing—chlorosulfonated polyethylene (Hypalon)	28.5	
—vinylidene fluoride/hexafluoropropylene (Fluorel, Viton A)	14.0–15.1	
Gasoline	46.8	43.7
Jet fuel—JP1		43.0
—JP3		43.5
—JP4	46.6	43.5
—JP5	45.9	43.0
Kerosene (jet fuel A)	46.4	43.3
Lanolin (wool fat)	40.8	
Lard	40.1	
Leather	18.2–19.8	
Lignin, $C_{2.6}H_3O$	24.7–26.4	23.4–25.1

Continued

Table 4.8   Heats of Combustion of Miscellaneous Materials *(continued)*

Materials	Gross, Δh_c^u (MJ/kg)	Net, Δh_c^l (MJ/kg)
Lignite	22.4–33.3	
Modacrylic fiber	24.7	
Naphtha	43.0–47.1	40.9–43.9
Neoprene, C_5H_5Cl—gum	24.3	
—foam	9.7–26.8	
Nomex™ (polymethaphenylene isophthalamide) fiber, $C_{14}H_{10}O_2N_2$	27.0–28.7	
Oil—castor	37.1	
—linseed	39.2–39.4	
—mineral	45.8–46.0	
—olive	39.6	
—solar	41.8	
Paper—brown	16.3–17.9	
—magazine	12.7	
—newsprint	19.7	
—wax	21.5	
Paraffin wax	46.2	43.1
Peat	16.7–21.6	
Petroleum jelly ($C_{7.118}H_{12.957}O_{0.091}$)	45.9	
Rayon fiber	13.6–19.5	
Rubber—buna N	34.7–35.6	
—butyl	45.8	
—isoprene (natural) C_5H_8	44.9	42.3
—latex foam	33.9–40.6	
—GRS	44.2	
—tire, auto	32.6	
Silicone rubber (SiC_2H_6O)	15.5–16.8	
—foam	14.0–19.5	
Sisal	15.9	
Spandex fiber	31.4	
Starch	17.6	16.2
Straw	15.6	
Sulfur—rhombic		9.28
—monoclinic		9.29
Tobacco	15.8	
Wheat	15.0	

Table continues below.

Table 4.8   Heats of Combustion of Miscellaneous Materials *(continued)*

Materials	Gross, Δh_c^u (MJ/kg)	Net, Δh_c^l (MJ/kg)
Wood—beech	20.0	18.7
—birch	20.0	18.7
—douglas fir	21.0	19.6
—maple	19.1	17.8
—red oak	20.2	18.7
—spruce	21.8	20.4
—white pine	19.2	17.8
—hardboard	19.9	
Woodflour	19.8	
Wool	20.7–26.6	

Source: Table C.4, *SFPE Handbook of Fire Protection Engineering,* 3rd edition. Courtesy of the Society of Fire Protection Engineers, Bethesda, MD.

Table 4.9 Properties of Metals

Metal	Properties at 20°C				Thermal Conductivity k (W/m·°C)									
	ρ (kg/m³)	c_p (kJ/kg·°C)	k (W/m·°C)	α (m²/s ×10⁵)	−100°C −148°F	0°C 32°F	100°C 212°F	200°C 392°F	300°C 572°F	400°C 752°F	600°C 1112°F	800°C 1472°F	1000°C 1832°F	1200°C 2192°F
Aluminum														
Pure	2,707	0.896	204	8.418	215	202	206	215	228	249				
Al-Cu (Duralumin) 94–96% Al, 3–5% CU, trace Mg	2,787	0.883	164	6.676	126	159	182	194						
Al-Si (Silumin, copper bearing) 86.5% Al,1% Cu	2,659	0.867	137	5.933	119	137	144	152	161					
Al-Si (Alusil) 78–80% Al, 20–22% Si	2,627	0.854	161	7.172	144	157	168	175	178					
Al-Mg-Si 97% Al, 1% Mg, 1% Si, 1% Mn	2,707	0.892	177	7.311		175	189	204						
Lead	11,373	0.130	35	2.343	36.9	35.1	33.4	31.5	29.8					
Iron														
Pure	7,897	0.452	73	2.034	87	73	67	62	55	48	40	36	35	36
Wrought iron 0.5% C	7,849	0.46	59	1.626		59	57	52	48	45	36	33	33	33
Steel (C max ≈ 1.5%):														
Carbon steel C ≈ 0.5%	7,833	0.465	54	1.474		55	52	48	45	42	35	31	29	31
1.0%	7,801	0.473	43	1.172		43	43	42	40	36	33	29	28	29
1.5%	7,753	0.486	36	0.970		36	36	36	35	33	31	28	28	29
Nickel steel														
Ni = 0%	7,897	0.452	73	2.026										
20%	7,933	0.46	19	0.526										
40%	8,169	0.46	10	0.279										
80%	8,618	0.46	35	0.872										
Invar 36% Ni	8,137	0.46	10.7	0.286										

Continued

Table 4.9 Properties of Metals (continued)

Metal	Properties at 20°C ρ (kg/m³)	c_p (kJ/kg·°C)	k (W/m·°C)	α (m²/s × 10⁵)	Thermal Conductivity k (W/m·°C) -100°C -148°F	0°C 32°F	100°C 212°F	200°C 392°F	300°C 572°F	400°C 752°F	600°C 1112°F	800°C 1472°F	1000°C 1832°F	1200°C 2192°F
Chrome steel														
Cr = 0%	7,897	0.452	73	2.026	87	73	67	62	55	48	40	36	35	36
1%	7,865	0.46	61	1.665		62	55	52	47	42	36	33	33	
5%	7,833	0.46	40	1.110		40	38	36	36	33	29	29	29	
20%	7,689	0.46	22	0.635		22	22	22	22	24	24	26	29	
Cr-Ni, chrome-nickel 15% Cr, 10% Ni	7,865	0.46	19	0.526										
18% Cr, 8% Ni (V2A)	7,817	0.46	16.3	0.444		16.3	17	17	19	19	19	22	26	31
20% Cr, 15% Ni	7,833	0.46	15.1	0.415										
25% Cr, 20% Ni	7,865	0.46	12.8	0.361										
Tungsten steel														
W = 0%	7,897	0.452	73	2.026										
1%	7,913	0.448	66	1.858										
5%	8,073	0.435	54	1.525										
10%	8,314	0.419	48	1.391										
Copper														
Pure	8,954	0.3831	386	11.234	407	386	379	374	369	363	353			
Aluminum bronze 95% Cu, 5% Al	8,666	0.410	83	2.330										
Bronze 75% Cu, 25% Sn	8,666	0.343	26	0.859										
Red brass 85% Cu, 9% Sn, 6% Zn	8,714	0.385	61	1.804	88	59	71							
Brass 70% Cu, 30% Zn	8,522	0.385	111	3.412			128	144	147	147				
German silver 62% Cu, 15% Ni, 22% Zn	8,618	0.394	24.9	0.733	19.2		31	40	45	48				

Continued

Table 4.9 Properties of Metals *(continued)*

Metal	Properties at 20°C				Thermal Conductivity k (W/m·°C)									
	ρ (kg/m³)	c_p (kJ/ kg·°C)	k (W/ m·°C)	α (m²/s × 10⁵)	-100°C -148°F	0°C 32°F	100°C 212°F	200°C 392°F	300°C 572°F	400°C 752°F	600°C 1112°F	800°C 1472°F	1000°C 1832°F	1200°C 2192°F
Constantan 60% Cu, 40% Ni	8,922	0.410	22.7	0.612	21		22.2	26						
Magnesium Pure	1,746	1.013	171	9.708	178	171	168	163	157					
Mg-Al (electrolytic) 6-8% Al, 1-2% Zn	1,810	1.00	66	3.605	138	52	62	74	83	109				
Molybdenum	10,220	0.251	123	4.790		125	118	114	111	109	106	102	99	92
Nickel Pure (99.9%)	8,906	0.4459	90	2.266	104	93	83	73	64	59				
Ni-Cr 90% Ni, 10% Cr	8,666	0.444	17	0.444		17.1	18.9	20.9	22.8	24.6				
80% Ni, 20% Cr	8,314	0.444	12.6	0.343		12.3	13.8	15.6	17.1	18.0	22.5			
Silver Purest	10,524	0.2340	419	17.004	419	417	415	412						
Pure (99.9%)	10,524	0.2340	407	16.563	419	410	415	374	362	360				
Tin, pure	7,304	0.2265	64	3.884	74	65.9	59	57						
Tungsten	19,350	0.1344	163	6.271		166	151	142	133	126	112	76		
Zinc, pure	7,144	0.3843	112.2	4.106	114	112	109	106	100	93				

Source: Table B.6, *SFPE Handbook of Fire Protection Engineering*, 3rd edition.
Courtesy of the Society of Fire Protection Engineers, Bethesda, MD.

Table 4.10 Melting, Boiling, and Ignition Temperatures of Pure Metals in Solid Forms

	Temperature					
	Melting Point		**Boiling Point**		**Solid Metal Ignition**	
Pure Metal	**°F**	**°C**	**°F**	**°C**	**°F**	**°C**
Aluminum	1220	660	4445	2452	1832[a,c]	1000[c]
Barium	1337	725	2084	1140	347[a]	175[a]
Calcium	1548	842	2625	1441	1300	704
Hafnium	4032	2222	9750	5399	—	—
Iron	2795	1535	5432	3000	1706[a]	930[a]
Lithium	367	186	2437	1336	356	180
Magnesium	1202	650	2030	1110	1153	623
Plutonium	1184	640	6000	3316	1112	600
Potassium	144	62	1400	760	156[a,b]	69[b,a]
Sodium	208	98	1616	880	239[c]	115[c]
Strontium	1425	774	2102	1150	1328[a]	720[a]
Thorium	3353	1845	8132	4500	932[a]	500[a]
Titanium	3140	1727	5900	3260	2900[a]	1593
Uranium	2070	1132	6900	3816	6900[a,d]	3816[a,d]
Zinc	786	419	1665	907	1652[a]	900[a]
Zirconium	3326	1830	6470	3577	2552[a]	1400[a]

[a]Ignition in oxygen.
[b]Spontaneous ignition in moist air.
[c]Above indicated temperature.
[d]Below indicated temperature.

Source: *Fire Protection Handbook,* 19th edition, Table 8.16.1.

INFORMATION FOR THE FIRE INVESTIGATOR
Properties of Materials

Table 4.11 Heats of Combustion for Metals

Material	Δh_c (MJ/kg)	Material	Δh_c (MJ/kg)
Pure elements		Copper alloys	
aluminum	31.04	bronze (88 Cu/10 Sb/2 Zn)	2.64
beryllium	2.46	red brass (85 Cu/15 Zn)	2.89
iron	4.87	cartridge brass (70 Cu/30 Zn)	3.33
magnesium	24.72	yellow brass (60 Cu/40 Zn)	3.62
manganese	7.01	Iron alloys	
molybdenum	6.13	carbon steels	7.4–7.5
nickel	4.10	stainless steels	7.7–8.4
tantalum	5.66	Nickel alloys	
tin	3.73	Incone 600	5.40
titanium	19.71	Monel 400	3.60
zinc	5.37		
zirconium	12.07		

Source: *Fire Protection Handbook,* 19th edition, Table A.4.

Table 4.12 Properties of Nonmetals

Substance	Temperature (°C)	k (W/m ·°C)	ρ (kg/m³)	C (kJ/kg·°C)	α (m²/s × 10⁷)
		Insulating material			
Asbestos					
Loosely packed	−45	0.149	470–570	0.816	3.3–4
	0	0.154			
	100	0.161			
Asbestos-cement boards	20	0.74			
Sheets	51	0.166			
Felt, 40 laminations/in.	38	0.057			
	150	0.069			
	260	0.083			
20 laminations/in.	38	0.078			
	150	0.095			
	260	0.112			
Corrugated, 4 plies/in.	38	0.087			
	93	0.100			
	150	0.119			
Asbestos cement	—	2.08			
Balsam wool, 2.2 lb/ft³	32	0.04	35		
Cardboard, corrugated	—	0.064			
Celotex	32	0.048			
Corkboard, 10 lb/ft³	30	0.043	160		
Cork, regranulated	32	0.045	45–120	1.88	2–5.3
Ground	32	0.043	150		
Diatomaceous earth (Sil-o-cel)	0	0.061	320		
Felt, hair	30	0.036	130–200		
Wool	30	0.052	330		
Fiber, insulating board	20	0.048	240		
Glass wool, 1.5 lb/ft³	23	0.038	24	0.7	22.6
Insulex, dry	32	0.064			
		0.144			
Kapok	30	0.035			
Magnesia, 85%	38	0.067	270		
	93	0.071			
	150	0.074			
	204	0.080			

Continued

INFORMATION FOR THE FIRE INVESTIGATOR
Properties of Materials

Table 4.12 Properties of Nonmetals *(continued)*

Substance	Temperature (°C)	k (W/m ·°C)	ρ (kg/m³)	C (kJ/kg·°C)	α (m²/s × 10⁷)
Insulating material *(contined)*					
Rock wool, 10 lb/ft³	32	0.040	160		
Loosely packed	150	0.067	64		
	260	0.087			
Sawdust	23	0.059			
Silica aerogel	32	0.024	140		
Wood shavings	23	0.059			
Structural and heat-resistant materials					
Asphalt	20–55	0.74–0.76			
Brick					
Building brick, common	20	0.69	1600	0.84	5.2
face		1.32	2000		
Carborundum brick	600	18.5			
	1400	11.1			
Chrome brick	200	2.32	3000	0.84	9.2
	550	2.47			9.8
	900	1.99			7.9
Diatomaceous earth, molded and fired	200	0.24			
	870	0.31			
Fireclay brick, burnt 2426°F	500	1.04	2000	0.96	5.4
	800	1.07			
	1100	1.09			
Fireclay brick, burnt 2642°F	500	1.28	2300	0.96	5.8
	800	1.37			
	1100	1.40			
Missouri	200	1.00	2600	0.96	4.0
	600	1.47			
	1400	1.77			
Magnesite	200	3.81		1.13	
	650	2.77			
	1200	1.90			

Table continues below.

Table 4.12 Properties of Nonmetals *(continued)*

Substance	Temperature (°C)	k (W/m·°C)	ρ (kg/m³)	C (kJ/kg·°C)	α (m²/s × 10⁷)
Insulating material *(continued)*					
Cement, portland		0.29	1500		
Mortar	23	1.16			
Concrete, cinder	23	0.76			
Stone 1-2-4 mix	20	1.37	1900–2300	0.88	8.2–6.8
Glass, window	20	0.78 (avg)	2700	0.84	3.4
Corosilicate	30–75	1.09	2200		
Plaster, gypsum	20	0.48	1440	0.84	4.0
Metal lath	20	0.47			
Wood lath	20	0.28			
Stone					
Granite		1.73–3.98	2640	0.82	8–18
Limestone	100–300	1.26–1.33	2500	0.90	5.6–5.9
Marble		2.07–2.94	2500–2700	0.80	10–13.6
Sandstone	40	1.83	2160–2300	0.71	11.2–11.9
Wood (across the grain)					
Balsa 8.8 lb/ft³	30	0.055	140		
Cypress	30	0.097	460		
Fir	23	0.11	420	2.72	0.96
Maple or oak	30	0.166	540	2.4	1.28
Yellow pine	23	0.147	640	2.8	0.82
White pine	30	0.112	430		

Source: Table B.7, *SFPE Handbook of Fire Protection Engineering,* 3rd edition. Courtesy of the Society of Fire Protection Engineers, Bethesda, MD.

Table 4.13 Fire Hazard Properties of Common Textile Fibers

Fiber Designation	Decomp. Temp. °F (°C)	Melting Temp. °F (°C)	Ignition Temp. °F (°C)	Burning Temp. °F (°C)	Burning Behavior	End Uses
A. *Natural Fibers*						
Cellulosic Cotton, hemp, jute, linen, sisal, etc.	580–610 (305–320)	—	490–750 (255–400)	1560 (850)	Char, burn, sometimes afterglow	Apparel, furnishings, towels, cordage
Protein Wool, mohair, cashmere, camel hair, etc.	450 (230)	—	1060–1110 (570–600)	1720 (940)	Char, intumesce, burn less readily than cellulosics	Apparel, blankets, carpets, furniture covers
B. *Manufactured Fibers*						
Acetate	570 (300)	500 (260)	842 (450)	1720 (940)	Shrinks, melts, and burns	Apparel, lingerie, furnishings
Acrylic	540–590 (280–310)	420–490 (215–255)	860–1050 (460–565)	1560 (850)	Chars, intumesces, burns, and drips	Apparel, furnishings, carpets, blankets, pile fabrics
Nylon	600–790 (315–420)	420–490 (215–255)	840–1060 (450–570)	1600 (870)	Melts, ablates, burnsuu	Apparel, lingerie, furnishings, carpets, cordage, industrial uses
Anidex	N/A	N/A	N/A	N/A	N/A	Elastic fiber used in apparel, home furnishing, and laces

Continued

Table 4.13 Fire Hazard Properties of Common Textile Fibers *(continued)*

Fiber Designation	Decomp. Temp. °F (°C)	Melting Temp. °F (°C)	Ignition Temp. °F (°C)	Burning Temp. °F (°C)	Burning Behavior	End Uses
Aramid	660 (370–400)	3700 (2040)	800–1040 (427–560)		Does not melt	
Lyocell	350 (175)	450–500 (230–260)	860 (460)			Dresses, slacks, coats, jeans
Modacrylic	500 (287)		1025 (550)			Children's sleepwear, carpets, blankets, draperies
Novoloid		Nonmelting				Protective equipment
Rubber	200 (93)					
Triacetate	>572 (>300)	550 (288)	>842 (>450)	1560 (850)	Shrinks, melts and burns	
Vinal	Similar to nylon					Good chemical resistance
Vinyon						Good chemical resistance
Olefin (polypropylene)	750 (400)	330 (165)	930–1060 (500–570)	1540 (840)	Melts, ablates, burns	Knitted sportswear, carpets, cordage, furniture covers, industrial uses

Continued

Table 4.13 Fire Hazard Properties of Common Textile Fibers (continued)

Fiber Designation	Decomp. Temp. °F (°C)	Melting Temp. °F (°C)	Ignition Temp. °F (°C)	Burning Temp. °F (°C)	Burning Behavior	End Uses
Polyester	680–750 (360–400)	480–570 (250–300)	840–1040 (450–560)	1290–1330 (700–720)	Melts, ablates, burns	Apparel, lingerie, furnishings, carpets, blankets, fiberfill, cordage, industrial uses
Rayon (viscose)	550 (290)	—	790 (420)	1560 (850)	Chars, burns	Apparel, lingerie, furnishings
Spandex	581–671 (305–355)	446–482 (230–250)	780 (415)	NA[a]	Melts, burns	In apparel and lingerie where stretch is desired

[a]NA Not available.
Source: *Fire Protection Handbook*, 19th edition, Table 8.5.5.

Tables 4.14 through 4.22 present selected properties of woods.

Table 4.14 Ignition Temperatures of Woods

Wood 2-1/2 in. (63.5 mm) long and Weighing 3 g (0.10 oz)	Unpiloted Ignition Temperature	
	°F	°C
Western red cedar	378	192
White pine	406	207
Long leaf pine	428	220
White oak	410	210
Paper birch	399	204

Source: *Fire Protection Handbook,* 19th edition, Table 8.3.2, from Brown, C. R., "The Ignition Temperatures of Solid Materials," *NFPA Quarterly,* Vol. 28, No. 2, 1934, pp. 134–135.

Table 4.15 Time Required to Ignite Wood Specimens

Wood 1-1/4 by 1-1/4 by 4 in. (32 × 32 × 102 mm)	No Ignition in 40 min		Exposure before Ignition, by Pilot Flame (min)						
	°F	°C	356°F (180°C)	392°F (200°C)	437°F (225°C)	482°F (250°C)	572°F (300°C)	662°F (350°C)	752°F (400°C)
Long leaf pine	315	157	14.3	11.8	8.7	6.0	2.3	1.4	0.5
Red oak	315	157	20.0	13.3	8.1	4.7	1.6	1.2	0.5
Tamarack	334	167	29.9	14.5	9.0	6.0	2.3	0.8	0.5
Western larch	315	157	30.8	25.0	17.0	9.5	3.5	1.5	0.5
Noble fir	369	187	—	—	15.8	9.3	2.3	1.2	0.3
Eastern hemlock	356	180	—	13.3	7.2	4.0	2.2	1.2	0.3
Redwood	315	157	28.5	18.5	10.4	6.0	1.9	0.8	0.3
Sitka spruce	315	157	40.0	19.6	8.3	5.3	2.1	1.0	0.3
Basswood	334	167	—	14.5	9.6	6.0	1.6	1.2	0.3

Source: *Fire Protection Handbook,* 19th edition, Table 8.3.4, from McNaughton, G. C., "Ignition and Charring Temperatures of Wood," Mimeo No. R1464, Forest Products Laboratory, Madison, WI 1960.

INFORMATION FOR THE FIRE INVESTIGATOR
Properties of Materials

Table 4.16 Ignition Temperatures of Wood Shavings

Wood Shavings	Unpiloted Ignition Temperature	
	°F	°C
Short leaf pine	442	228
Long leaf pine	446	230
Douglas fir	500	260
Spruce	502	261
White pine	507	264

Source: *Fire Protection Handbook,* 19th edition, Table 8.3.3.

Table 4.17 Flame Spread Indexes for Selected Wood Species, 1-in. (25.4 mm) Nominal Solid Dimension Lumber

Species	Flame Spread Index	Source
Bald cypress	145–150	UL[a]
Eastern red cedar	110	HUD/FHA[b]
Alaska yellow cedar	78	CWC[c]
Western red cedar	70	HPMA[d]
Douglas fir	70–100	UL
Western hemlock	60–75	UL
Western white pine	75, 72	UL, HPMA
Eastern white pine	120–215, 85	UL, CWC
Lodgepole pine	65–110	CWC
Ponderosa pine	105–230	UL
Red pine	142	CWC
Southern yellow pine	130–195	UL
Redwood	70	UL
Eastern white spruce	65	CWC, UL
Sitka spruce	100	UL
Yellow birch	105–110	UL
Cottonwood	115	UL
White (sugar) maple	104	CWC
Red oak	100	UL
White oak	100	UL
Sweetgum	140–155	UL
Walnut	130–140	UL
Yellow poplar	170–185	UL

[a]UL: Underwriters Laboratories Inc.

[b]HUD/FHA: U.S. Dept. of Housing and Urban Development.

[c]CWC: Canadian Wood Council.

[d]HPMA: Hardwood Plywood Manufacturers Association.

Note: Flame spread ratings developed for virtually all wood-based panel products are based upon ASTM E84, UL 723, or NFPA 255 flame spread rating tests, using the Steiner Test Tunnel.

Source: *Fire Protection Handbook,* 19th edition, Table 8.3.5, from Wood Handbook: Wood as an Engineering Material, USDA Forest Service Handbook, No. 72, Government Printing Office, Washington, DC, 1987.

INFORMATION FOR THE FIRE INVESTIGATOR
Properties of Materials

Table 4.18 ASTM E119 Fire Resistance Data for Selected Woods

Species	Specific Gravity	Rate (in./hr)	Rate mm/hr
Bald cypress	0.44	1.7	43.2
Basswood	0.42	2.4	61.0
Yellow birch	0.63	2.0	50.8
Chestnut	0.45	1.7	43.2
Douglas fir	0.45	1.6	40.6
Eastern hemlock	0.40	1.6	40.6
White (sugar) maple	0.64	2.1	53.3
Northern red oak	0.61	1.8	45.7
White oak	0.67	1.5	38.1
Eastern white pine	0.39	1.6	40.6
Ponderosa pine	0.42	2.1	53.3
Southern yellow pine	0.55	2.2	55.9
Sugar pine	0.32	2.0	50.8
Redwood	0.38	1.6	40.6
Sitka spruce	0.43	1.7	43.2
Sweetgum-sapwood	0.55	2.4	61.0
Sweetgum-heartwood	0.52	1.6	40.6
Yellow poplar	0.44	2.1	53.3

Source: *Fire Protection Handbook,* 19th edition, Table 8.3.6, from Wood Handbook: Wood as an Engineering Material, USDA Forest Service Handbook, No. 72, Government Printing Office, Washington, DC, 1987.

Table 4.19 Surface Burning Characteristics of Pressure-Treated and Untreated Wood[a]

Species	FSI[b] (untreated)	FSI[b] Pressure-Impregnated	Smoke-Developed Pressure Impregnation
Douglas fir (lumber)	70–100	≤25	≤25
Hemlock (lumber)	60–70	≤25	≤25
Southern yellow pine (lumber)	130–195	≤25	≤25
Western red cedar (lumber)	70	≤25	≤25
Redwood (lumber)	65–70	≤25	≤25
Red oak (lumber)	91	≤25	≤25
Western spruce (lumber)	100	≤25	≤25
Douglas fir (plywood)	115–130	≤25	≤25
Southern yellow pine (plywood)	95	≤25	≤25
Particleboard	140–200	≤25	≤25

[a]As tested by NFPA 255 (ASTM E84), selected species only.
[b]Flame spread index.

Source: *Fire Protection Handbook,* 19th edition, Table 8.4.1, from Wood Handbook: Wood as an Engineering Material, USDA Forest Service Handbook, No. 72, Government Printing Office, Washington, DC, 1987.

Table 4.20 Heat Release Rates of Pressure-Treated and Untreated Wood[a]

Species (Product)	Irradiance (kW/m²)	Average Peak HRR[b](kW)	Average HRR (kW)
Southern yellow pine (plywood)[a]	50	237	156
Southern yellow pine (fire-retardant treated plywood)[a]	50	151	55

[a]Unpublished data generated at Weyerhaeuser Company.
[b]HRR—Heat release rate.
Source: *Fire Protection Handbook,* 19th edition, Table 8.4.2.

Table 4.21 Effect of Pressure Impregnation
on Douglas Fir Plywood in the Room Fire Test[a]

	Untreated	Pressure Impregnated
Flashover time	3.3 min	None
Maximum heat release rate	7.3 mW	0.6 mW
Maximum smoke generation (O.D.[b])	1.11	0.76

[a]The ignition exposure was 40 kW (net) for 5 min. 160 kW (net) for 5 min, and 0 kW (net) for 5 min.
[b]Optical density.
Source: *Fire Protection Handbook,* 19th edition, Table 8.4.3.

Table 4.22 Variations in Effect of Different
Fire-Retardant Coatings on Unprimed Douglas Fir

	Untreated	Painted with Coating X[a]	Painted with Coating Y[a]	Painted with Coating Z[a]
Total coverage	—	100	100	200
[sq ft per gal (L/m^2)]	—	2.45 L/m^2	2.45 L/m^2	1.23 L/m^2
Flame spread[b]	100	60	10	25
Fuel contributed[b]	100	35	15	10
Smoke developed[b]	100	40	5	10
Rate per coat	—	300	200	200
[sq ft per gal (L/m^2)]	—	0.82 L/m^2	1.23 L/m^2	1.23 L/m^2
Number of coats	—	3	2	1

[a]Each coating is listed by Underwriters Laboratories, Inc., as meeting the minimum requirements for a fire-retardant coating.
[b]Where a range of values was obtained, the figure given here is the highest obtained.
Source: *Fire Protection Handbook,* 19th edition, Table 8.4.4.

Properties of Liquids and Gases

Tables 4.23 through 4.27 present selected properties of liquids and gases.

Table 4.23 Fire Hazard Properties of Some Common Liquids

	Flash Point °F (°C)	Ignition Temp. °F (°C)	Flammable Limits Percent by Vol. Lower	Upper	Sp. Gr. (Water =1)	Vapor Density (Air =1)	Boiling Point °F (°C)	Water Soluble	See Introduction for Suggested Extinguishing Methods	Hazard Identification Health	Flammability	Reactivity
Acetaldehyde CH_3CHO (Acetic Aldehyde) (Ethanol)	-38 (-39)	347 (175)	4.0	60	0.8	1.5	70 (21)	Yes	1	3	4	2
Note: Polymerizes. See Hazardous Chemicals Data.												
Acetone CH_3COCH_3 (Dimethyl Ketone) (2-Propanone)	-4 (-20)	869 (465)	2.5	12.8	0.8	2.0	133 (56)	Yes	1	1	3	0
Acetyl Chloride CH_3COCl (Ethanoyl Chloride)	40 (4)	734 (390)			1.1	2.7	124 (51)	Violent decomposition	5	3	3	2W
Note: See Hazardous Chemicals Data.												
Acrolein Dimer $(CH_2{:}CHCHO)_2$	118 (48) (oc)				1.1		304 (151)	Yes	5	1	2	1
Note: See Hazardous Chemicals Data.												
Acrylonitrile $CH_2{:}CHCN$ (Vinyl Cyanide) (Propenenitrile)	32 (0) (oc)	898 (481)	3.0	17	0.8	1.8	171 (77)	Yes	1	4	3	2
Note: Polymerizes. See Hazardous Chemicals Data.												
Allyl Alcohol $CH_2{:}CHCH_2OH$	70 (21)	713 (378)	2.5	18.0	0.9	2.0	206 (97)	Yes	1	4	3	1
Note: See Hazardous Chemicals Data.												

Continued

Table 4.23 Fire Hazard Properties of Some Common Liquids (continued)

	Flash Point °F (°C)	Ignition Temp. °F (°C)	Flammable Limits Percent by Vol. Lower	Flammable Limits Percent by Vol. Upper	Sp. Gr. (Water = 1)	Vapor Density (Air = 1)	Boiling Point °F (°C)	Water Soluble	See Introduction for Suggested Extinguishing Methods	Hazard Identification Health	Hazard Identification Flammability	Hazard Identification Reactivity
Allylamine $CH_2{:}CHCH_2NH_2$ (2-Propenylamine)	-20 (-29)	705 (374)	2.2	22	0.8	2.0	128 (53)	Yes	1 5	4	3	1
Note: See Hazardous Chemicals Data.												
Amyl Acetate $CH_3COOC_5H_{11}$ (1-Pentanol Acetate) Commercial Grade	60 (16) 70 (21)	680 (360)	1.1	7.5	0.9	4.5	300 (149)	Slight	1 5	1	3	0
Amyl Alcohol $CH_3(CH_2)_3CH_2OH$ (1-Pentanol)	91 (33)	572 (300)	1.2	10.0 @ 212 (100)	0.8	3.0	280 (138)	Slight	5	1	3	0
Asphalt (Typical) (Petroleum Pitch)	400+ (204+)	905 (485)			1.0-1.1	2.8	>700 (>371)	No	2	0	1	0
Benzene C_6H_6 (Benzol)	12 (-11)	928 (498)	1.2	7.8	0.9	2.8	176 (80)	No	1	2	3	0
Note: See Hazardous Chemicals Data.												
Benzyl Alcohol $C_6H_5CH_2OH$ (Phenyl Carbinol)	200 (93)	817 (436)			1.0+		403 (206)	Slight	5 2	2	1	0

Continued

Table 4.23 Fire Hazard Properties of Some Common Liquids (continued)

	Flash Point °F (°C)	Ignition Temp. °F (°C)	Flammable Limits Percent by Vol. Lower	Flammable Limits Percent by Vol. Upper	Sp. Gr. (Water = 1)	Vapor Density (Air = 1)	Boiling Point °F (°C)	Water Soluble	See Introduction for Suggested Extinguishing Methods	Hazard Identification Health	Hazard Identification Flammability	Hazard Identification Reactivity
Butyl Acetate CH₃COOC₄H₉ (Butylethanoate)	72 (22)	797 (425)	1.7	7.6	0.9	4.0	260 (127)	Slight	1 5	1	3	0
Butyl Acrylate CH₂:CHCOOC₄H₉	84 (29)	559 (292)	1.7	9.9	0.9	4.4	260 (127)	No		2	2	2
Butyl Alcohol CH₃(CH₂)₂CH₂OH (1-Butanol) (Propylcarbinol) (Propyl Methanol)	98 (37)	650 (343)	1.4	11.2	0.8	2.6	243 (117) Polymerizes	No	1 5	1	3	0
Carbon Disulfide CS₂ (Carbon Bisulfide)	–22 (–30)	194 (90)	1.3	50.0	1.3	2.6	115 (46)	No	4	3	3	0
	Note: See Hazardous Chemicals Data.											
Chlorobenzene C₆H₅Cl (Chlorobenzol) (Monochlorobenzene) (Phenyl Chloride)	82 (28)	1099 (593)	1.3	9.6	1.1	3.9	270 (132)	No	4	2	3	0
	Note: See Hazardous Chemicals Data.											
Corn Oil	490 (254)	740 (393)			0.9			No	2			

Continued

Table 4.23 Fire Hazard Properties of Some Common Liquids *(continued)*

	Flash Point °F (°C)	Ignition Temp. °F (°C)	Flammable Limits Percent by Vol. Lower	Flammable Limits Percent by Vol. Upper	Sp. Gr. (Water = 1)	Vapor Density (Air = 1)	Boiling Point °F (°C)	Water Soluble	See Introduction for Suggested Extinguishing Methods	Hazard Identification Health	Hazard Identification Flammability	Hazard Identification Reactivity
Corn Oil *(continued)*												
Cooking	610 (321) (0c)				<1				2	0	1	0
Creosote Oil	165 (74)	637 (336)			>1			No	3	2	2	0
Cumene $C_6H_5CH(CH_3)_2$ (Cumol) (2-Phenyl Propane) (Isopropyl Benzene)	96 (36)	795 (424)	0.9	6.5	0.9	4.1	382–752 (194–400)	No	3	2	3	1
Cyclohexane C_6H_{12} (Hexahydrobenzene) (Hexamethylene)	−4 (−20)	473 (245)	1.3	8	0.8	2.9	179 (82)	No	1	1	3	0
Denatured Alcohol	60 (16)	750 (399)			0.8	1.6	175 (79)	Yes	1 5	0	3	0
Government Formula												
CD-5	60–62 (16–17)											
CD-5A	60–61 (15.5–16)											
CD-10	49–59 (9–15)											

Continued

Table 4.23 Fire Hazard Properties of Some Common Liquids (continued)

	Flash Point °F (°C)	Ignition Temp. °F (°C)	Flammable Limits Percent by Vol. Lower	Flammable Limits Percent by Vol. Upper	Sp. Gr. (Water = 1)	Vapor Density (Air = 1)	Boiling Point °F (°C)	Water Soluble	See Introduction for Suggested Extinguishing Methods	Hazard Identification Health	Hazard Identification Flammability	Hazard Identification Reactivity
Denatured alcohol (continued)												
SD-1	57 (14)											
SD-2B	56 (13)											
SD-3A	59 (15)											
SD-13A	<19 (<−7)											
SD-17	60 (16)											
SD-23A	35 (2)											
SD-30	59 (15)											
SD-39B	60 (16)											
SD-39C	59 (15)											
SD-40M	59 (15)											
Dibutylamine ($C_4H_9)_2NH$	117 (47)		1.1	0.8	4.5	322 (161)	Slight		5	3	2	0
Dibutyl Ether ($C_4H_9)_2O$ (1-Butoxybutane) (Butyl Ether)	77 (25)	382 (194)	1.5	7.6	0.8	4.5	286 (141)	No	1	2	3	1

Note: See Hazardous Chemicals Data.

Continued

Table 4.23 Fire Hazard Properties of Some Common Liquids (continued)

	Flash Point °F (°C)	Ignition Temp. °F (°C)	Flammable Limits Percent by Vol. Lower	Upper	Sp. Gr. (Water = 1)	Vapor Density (Air = 1)	Boiling Point °F (°C)	Water Soluble	See Introduction for Suggested Extinguishing Methods	Hazard Identification Health	Flammability	Reactivity
Diesel Fuel Oil No. 1-D	100 Min. (38) or legal									0	2	0
Diesel Fuel Oil No. 2-D	125 Min. (52) or legal									0	2	0
Diethylamine $(C_2H_5)_2NH$	-9 (-23)	594 (312)	1.8	10.1	0.7	2.5	134 (57)	Yes	5 1	3	3	0
Note: See Hazardous Chemicals Data.												
Epichlorohydrin CH_2CHOCH_2Cl (2-Chloropropylene Oxide) (γ-Chloropropylene Oxide)	88 (31)	772 (411)	3.8	21.0	1.2	3.2	239 (115)	Yes	5	3	3	2
Note: See Hazardous Chemicals Data.												
Ethyl Acetate $CH_3COOC_2H_5$ (Acetic Ester) (Acetic Ether) (Ethyl Ethanoate)	24 (-4)	800 (426)	2.0	11.5	0.9	3.0	171 (77)	Slight	1 5	1	3	0
Ethyl Acrylate $CH_2{:}CHCOOC_2H_5$	50 (10) (oc)	702 (372)	1.4	14	0.9	3.5	211 (99)	Slight	1 5	2	3	2
Note: Polymerizes. See Hazardous Chemicals Data.												

Continued

Table 4.23 Fire Hazard Properties of Some Common Liquids (continued)

	Flash Point °F (°C)	Ignition Temp. °F (°C)	Flammable Limits Percent by Vol. Lower	Upper	Sp. Gr. (Water = 1)	Vapor Density (Air = 1)	Boiling Point °F (°C)	Water Soluble	See Introduction for Suggested Extinguishing Methods	Health	Flammability	Reactivity
Ethyl Alcohol C₂H₅OH (Grain Alcohol, Cologne Spirits, Ethanol)	55 (13)	685 (363)	3.3	19	0.8	1.6	173 (78)	Yes	1 5	0	3	0
Ethyl Alcohol and Water												
96%	62 (17)											
95%	63 (17)											
80%	68 (20)											
70%	70 (21)											
60%	72 (22)											
50%	75 (24)											
40%	79 (26)											
30%	85 (29)											
20%	97 (36)											
10%	120 (49)											
5%	144 (62)											

Continued

Table 4.23 Fire Hazard Properties of Some Common Liquids *(continued)*

	Flash Point °F (°C)	Ignition Temp. °F (°C)	Flammable Limits Percent by Vol.		Sp. Gr. (Water = 1)	Vapor Density (Air = 1)	Boiling Point °F (°C)	Water Soluble	See Introduction for Suggested Extinguishing Methods	Hazard Identification		
			Lower	Upper						Health	Flammability	Reactivity
Ethylamine $C_2H_5NH_2$ 70% Aqueous Solution (Aminoethane)	<0 (<-18)	725 (385)	3.5	14.0	0.8	1.6	62 (17)	Yes	1 5	3	4	0
Note: See Hazardous Chemicals Data.												
Ethyl Chloride C_2H_5Cl (Chloroethane) (Hydrochloric Ether) (Muriatic Ether)	-58 (-50)	966 (519)	3.8	15.4	0.9	2.2	54 (12)	Slight	1	1	4	0
Note: See Hazardous Chemicals Data.												
Ethylene Dichloride CH_2ClCH_2Cl (1,2 Dichloroethane) (Glycol Dichloride)	56 (13)	775 (413)	6.2	16	1.3	3.4	183 (84)	No	4	2	3	0
Note: See Hazardous Chemicals Data.												
Ethylene Oxide CH_2OCH_2 (Dimethylene Oxide) (1,2-Epoxyethane) (Oxirane)	-20	1058 with No Air	3.0	100	0.9	1.5	51 (11)	Yes	1	3 Vapors Explosive.	4	3
Note: See Hazardous Chemicals Data.												
Ethyl Ether $C_2H_5OC_2H_5$ (Diethyl Ether)	-49 (-45)	356 (180)	1.9	36.0	0.7	2.6	95 (35)	Slight	1 5	1	4	1

Continued

Table 4.23 Fire Hazard Properties of Some Common Liquids (continued)

	Flash Point °F (°C)	Ignition Temp. °F (°C)	Flammable Limits Percent by Vol. Lower	Upper	Sp. Gr. (Water = 1)	Vapor Density (Air = 1)	Boiling Point °F (°C)	Water Soluble	See Introduction for Suggested Extinguishing Methods	Hazard Identification Health	Flammability	Reactivity
Ethyl Ether *(continued)*												
(Diethyl Oxide)												
(Ether)												
(Ethyl Oxide)												
								Note: See Hazardous Chemicals Data.				
Ethyl Methacrylate	68				0.9	3.9	239–248	No	1	2	3	0
$CH_2{:}C(CH_3)COOC_2H_5$	(20)						(115–120)					
(Ethyl Methyl Acrylate)	(oc)											
Fuel Oil No. 1	100–162	410	0.7	5	<1		304–574	No	0	0	2	0
(Kerosene)	(38–72)	(210)					(151–301)					
(Range Oil)												
(Coal Oil)												
Fuel Oil No. 2	126–204	494			<1			No	0	0	2	0
	(52–96)	(257)										
Fuel Oil No. 4	142–240	505			<1			No	0	0	2	0
	(61–116)	(263)										
Fuel Oil No. 6	150–270	765			1±			No	1	0	2	0
	(66–132)	(407)										
Gasoline	–45	536	1.4	7.6	0.8	3–4	100–400	No	1	1	3	0
C_5H_{12} to C_9H_{20}	(–43)	(280)	1.4	7.6			(38–204)					
56–60 octane	–45											
	(–43)											

Continued

Table 4.23 Fire Hazard Properties of Some Common Liquids (continued)

	Flash Point °F (°C)	Ignition Temp. °F (°C)	Flammable Limits Percent by Vol.		Sp. Gr. (Water = 1)	Vapor Density (Air = 1)	Boiling Point °F (°C)	Water Soluble	See Introduction for Suggested Extinguishing Methods	Hazard Identification		
			Lower	Upper						Health	Flammability	Reactivity
Gasoline (cont'd)												
73 Octane			1.4	7.6								
92 Octane			1.5	7.6								
100 Octane	−36 (−38)	853 (456)	1.4	7.4								
Note: Values may vary considerably for different grades of gasoline.												
Gasoline 100–130 Octane (Aviation Grade)	−50 (−46) approx.	824 (440)	1.3	7.1						1	3	0
Gasoline 115–145 Octane (Aviation Grade)	−50 (−45) approx.	880 (471)	1.2	7.1						1	3	0
Hexane CH$_3$(CH$_2$)$_4$CH$_3$ (Hexyl Hydride)	−7 (−22)	437 (225)	1.1	7.5	0.7	3.0	156 (69)	No	1	1	3	0
Isopropyl Alcohol (CH$_3$)$_2$CHOH (Isopropanol) (Dimethyl Carbinol) (2-Propanol) 87.9% iso	53 (12) 57 (14)	750 (399)	2.0	12.7 @ 200 (93)	0.8	2.1	181 (83)	Yes	5 1	1	3	0

Continued

Table 4.23 Fire Hazard Properties of Some Common Liquids *(continued)*

	Flash Point °F (°C)	Ignition Temp. °F (°C)	Flammable Limits Percent by Vol. Lower	Upper	Sp. Gr. (Water = 1)	Vapor Density (Air = 1)	Boiling Point °F (°C)	Water Soluble	See Introduction for Suggested Extinguishing Methods	Health	Flammability	Reactivity
Jet Fuels Jet A and Jet A-1	110–150 (43–66)						400–550 (204–288)		0	2		0
Jet Fuels Jet B	−10 to +30 (−23 to −1)											
Kerosene	See Fuel Oil No. 1								1	1	3	0
Linseed Oil, Raw Boiled	432 (222) 403 (206)	650 (343)			0.9		600+ (316+)	No	2	0	1	0
Methyl Alcohol CH₃OH (Methanol) (Wood Alcohol) (Columbian Spirits)	52 (11)	867 (464)	6.0	36	0.8	1.1	147 (64)	Yes	1	1	3	0
Methylamine CH₃NH₂	Gas	806 (430)	4.9	20.7		1.0	21 (−6)	Yes	6	3	4	0

Note: See Hazardous Chemicals Data.

INFORMATION FOR THE FIRE INVESTIGATOR
Properties of Materials

Continued

Table 4.23 Fire Hazard Properties of Some Common Liquids (continued)

	Flash Point °F (°C)	Ignition Temp. °F (°C)	Flammable Limits Percent by Vol. Lower	Upper	Sp. Gr. (Water = 1)	Vapor Density (Air = 1)	Boiling Point °F (°C)	Water Soluble	See Introduction for Suggested Extinguishing Methods	Hazard Identification Health	Flammability	Reactivity
Methyl Chloride CH₃Cl (Chloromethane)	-50	1170 (632)	8.1	17.4		1.8	-11 (-24)	Slight	6	1	4	0
Note: See Hazardous Chemicals Data.												
Methyl Ether (CH₃)₂O (Dimethyl Ether) (Methyl Oxide)	Gas	662 (350)	3.4	27.0		1.6	-11 (-24)	Yes	6	1	4	1
Methyl Ethyl Ketone C₂H₅COCH₃ (2-Butanone) (Ethyl Methyl Ketone)	16 (-9)	759 (404)	1.4 @ 200 (93)	11.4 @ 200 (93)	0.8	2.5	176 (80)	Yes	1 5	1	3	0
Methyl Methacrylate CH₂:C(CH₃)COOCH₃	50 (10) (oc)		1.7	8.2	0.9	3.6	212 (100)	Very slight	1	2	3	2
Note: Polymerizes. See Hazardous Chemicals Data.												
Mineral Oil	380 (193) (oc)				0.8— 0.9		680 (360)	No	2	0	1	0
Naphtha, Petroleum	See Petroleum Ether.											

Continued

Table 4.23 Fire Hazard Properties of Some Common Liquids (continued)

	Flash Point °F (°C)	Ignition Temp. °F (°C)	Flammable Limits Percent by Vol. Lower	Upper	Sp. Gr. (Water = 1)	Vapor Density (Air = 1)	Boiling Point °F (°C)	Water Soluble	See Introduction for Suggested Extinguishing Methods	Health	Flammability	Reactivity
Naphtha V.M. & P., 50° Flash (10)	50 (10)	450 (232)	0.9	6.7	<1	4.1	240–290 (116–143)	No	1	1	3	0
Note: Flash point and ignition temperature will vary depending on the manufacturer.												
Naphtha, V.M. & P., High Flash	85 (29)	450 (232)	1.0	6.0	<1	4.3	280–350 (138–177)	No	1	1	3	0
Note: Flash point and ignition temperature will vary depending on the manufacturer.												
Naphtha V. M. & P., Regular	28 (−2)	450 (232)	0.9	6.0	<1	4.1	212–320 (100–160)	No	1	1	3	0
Note: Flash point and ignition temperature will vary depending on the manufacturer.												
Nitrobenzene $C_6H_5NO_2$ (Nitrobenzol) (Oil or Mirbane) See Hazardous Chemicals Data.	190 (88)	900 (482) @ 200 (93)	1.8	1.2	1.2	4.3	412 (211)	No	3	3	2	1
Nitromethane CH_3NO_2	95 (35)	785 (418)	7.3	1.1	1.1	2.1	214 (101)	Slight	1 · 5	1	3	4
Note: May detonate under high temperature and pressure conditions. See Hazardous Chemicals Data.												
Petroleum, Crude, Sweet	20–90 (−7 to 32)				<1			No	1	1	3	0

Continued

Table 4.23 Fire Hazard Properties of Some Common Liquids (continued)

	Flash Point °F (°C)	Ignition Temp. °F (°C)	Flammable Limits Percent by Vol. Lower	Upper	Sp. Gr. (Water = 1)	Vapor Density (Air = 1)	Boiling Point °F (°C)	Water Soluble	See Introduction for Suggested Extinguishing Methods	Hazard Identification Health	Flammability	Reactivity
Petroleum Ether (Benzine) (Naphtha, Petroleum)	<0 (<–18)	550 (288)	1.1	5.9	0.6	2.5	95–140 (35–60)	No	1	1	4	0
Phenol C_6H_5OH (Carbolic Acid)	175 (79)	1319 (715)	1.8	8.6	1.1	3.2	358 (181)	Yes	5	4	2	0
Note: See Hazardous Chemicals Data. Melting point 108 (42).												
Propyl Alcohol $CH_3CH_2CH_2OH$ (1-Propanol)	74 (23)	775 (412)	2.2	13.7	0.8	2.1	207 (97)	Yes	1 5	1	3	0
Propylene Oxide OCH_2CHCH_3	–35 (–37)	840 (449)	2.3	36	0.83	2.0	94 (35)	Yes	1 5	3	4	2
Note: See Hazardous Chemicals Data.												
Styrene $C_6H_5CH:CH_2$ (Cinnamene) (Phenylethylene) (Vinyl Benzene)	88 (31)	914 (490)	0.9	6.8	0.9	3.6	295 (146)	No	1	2	3	2
Note: Polymerizes. See Hazardous Chemicals Data.												
Tetrahydrofuran $OCH_2CH_2CH_2CH_2$ (Diethylene Oxide) (Tetramethylene Oxide)	6 (–14)	610 (321)	2	11.8	0.9	2.5	151 (66)	Yes	1 5	2	3	1
Note: See Hazardous Chemicals Data.												

Continued

Table 4.23 Fire Hazard Properties of Some Common Liquids (continued)

	Flash Point °F (°C)	Ignition Temp. °F (°C)	Flammable Limits Percent by Vol.		Sp. Gr. (Water = 1)	Vapor Density (Air = 1)	Boiling Point °F (°C)	Water Soluble	See Introduction for Suggested Extinguishing Methods	Hazard Identification		
			Lower	Upper						Health	Flammability	Reactivity
Toluene	40	896	1.1	7.1	0.9	3.1	231	No	1	2	3	0
C₆H₅CH₃	(4)	(480)					(111)					
(Methylbenzene)												
(Phenylmethane)												
(Toluol)												
Note: See Hazardous Chemical Data.												
Turpentine	95	488	0.8	<1			300	No	1	1	3	0
	(35)	(253)					(149)					
Note: See Hazardous Chemicals Data.												
Vinyl Acetate	18	756	2.6	13.4	0.9	3.0	161	Slight	1	2	3	2
CH₂:CHOOCCH₃	(−8)	(402)					(72)		5			
(Ethenyl Ethanoate)												
Note: Polymerizes. See Hazardous Chemicals Data.												
o-Xylene	90	867	0.9	6.7	0.9	3.7	292	No	1	2	3	0
C₆H₄(CH₃)₂	(32)	(463)					(144)					
(1,2-Dimethylbenzene)												
(o-Xylol)												
Note: See Hazardous Chemicals Data.												

Source: *Flammable and Combustible Liquids Code Handbook*, 6th edition, 1997, pp. 489–502.

INFORMATION FOR THE FIRE INVESTIGATOR
Properties of Materials

Table 4.24 Common Liquids That Have Low Conductivity

Typical Conductivity Product	Conductance per Meter in Pico-Siemen[a]
Highly purified hydrocarbons[b]	0.01
Light distillates[b]	0.01 to 10
Commercial jet fuel[c]	0.2 to 50
Kerosene[c]	1 to 50
Leaded gasoline[c]	above 50
Fuel with antistatic additives[c]	50 to 300
Black oils[b]	1000 to 100,000

[a]Pico-siemen is the reciprocal of ohms. One pico-siemen is 1 trillionth (1×10^{-12}) of a siemen.

[b]API RP 2003, *Protection Against Ignitions Arising Out of Static, Lightning, and Stray Currents.*

[c]Bustin and Duket, *Electrostatic Hazards in Petroleum Industry.*

Source: NFPA 921, 2001 edition, Table 6.12.2.1.

Table 4.25 Properties of Typical Flammable Gases

Flammable Gas	Molecular Weight	Btu/ft³	Autoignition (°F)	LEL% by Volume	UEL% by Volume	Vapor Density (Air = 1)	ft³ Air Req'd to Burn 1 ft³ of Gas
Butane	58	3200	550	1.9	8.5	2	31
CO	28	310	1128	12.5	74	0.97	2.5
Hydrogen	2	311	932	4	74.2	0.07	2.5
Natural gas (high Btu type)	18.6	1115	—	4.6	14.5	0.64	10.6
Natural gas (high methane type)	16.2	960	—	4	15	0.56	9
Natural gas (high inert type)	20.3	1000	—	3.9	14	0.70	9.4
Propane	44	2500	842	2.1	9.5	1.57	24

Source: NFPA 86, 1999 edition, Table A-4-2-1-1.

Table 4.26 Combustion Properties of Common Flammable Gases

Gas	Btu per cu ft (Gross)	mJ/m³ (Gross)	Limits of Flammability Percent by Volume in Air — Lower	Limits of Flammability Percent by Volume in Air — Upper	Specific Gravity (Air = 1.0)	Volume of Air Needed to Burn 1 cu ft of Gas (cu ft)	Volume of Air Needed to Burn 1 m³ of Gas (m³)	Ignition Temperature °F	Ignition Temperature °C
Natural gas									
High inert type[a]	958–1051	35.7–39.2	4.5	14.0	.660—.708	9.2	9.2	—	—
High methane type[b]	1008–1071	37.6–39.9	4.7	15.0	.590—.614	10.2	10.2	900–1170	482–632
High Btu type[c]	1071–1124	39.9–41.9	4.7	14.5	.620—.719	9.4	9.4	—	—
Blast furnace gas	81–111	3.0–4.1	33.2	71.3	1.04–1.00	0.8	0.8	—	—
Coke oven gas	575	21.4	4.4	34.0	.38	4.7	4.7	—	—
Propane (commercial)	2516	93.7	2.15	9.6	1.52	24.0	24.0	920–1120	493–604
Butane (commercial)	3300	122.9	1.9	8.5	2.0	31.0	31.0	900–1000	482–538
Sewage gas	670	24.9	6.0	17.0	0.79	6.5	6.5	—	—
Acetylene	1499	55.8	2.5	81.0	0.91	11.9	11.9	581	305
Hydrogen	325	12.1	4.0	75.0	0.07	2.4	2.4	932	500
Anhydrous ammonia	386	14.4	16.0	25.0	0.60	8.3	8.3	1204	651
Carbon monoxide	314	11.7	12.5	74.0	0.97	2.4	2.4	1128	609
Ethylene	1600	59.6	2.7	36.0	0.98	14.3	14.3	914	490
Methylacetylene, propadiene, stabilized[d]	2450	91.3	3.4	10.8	1.48	—	—	850	454

[a]Typical composition CH₄ 71.9–83.2%; N₂ 6.3–16.2%.
[b]Typical composition CH₄ 87.6–95.7%; N₂ 0.1–2.39%.
[c]Typical composition CH₄ 85.0–90.1%; N₂ 1.2–7.5%.
[d]MAPP® Gas.

Source: *Fire Protection Handbook*, 19th edition, Table 8.7.3.

Table 4.27 Approximate Properties of LP-Gases

	Commercial Propane	Commercial Butane
Vapor pressure in psig at		
70°F	127	17
100°F	196	37
105°F	210	41
130°F	287	69
Vapor pressure in kPa at		
20°C	895	103
40°C	1,482	285
45°C	1,672	345
55°C	1,980	462
Specific gravity of liquid at 60°F (15.5°C)	0.509	0.582
Initial boiling point at 14.7 psia	−44°F	15°F
Initial boiling point at 101 kPa	−42°C	−9°C
Weight per gal of liquid at 60°F, lb	4.20	4.81
Weight per m^3 of liquid at 15.5°C, kg	504 kg	582 kg
Specific heat of liquid, Btu/lb 60°F	0.630	0.549
Specific heat of liquid, kJ/kg at 15.5°C	1.46	1.28
Cu ft of vapor per gal at 60°F	36.38	31.26
m^3 of vapor per L at 15.5°C	0.271	0.235
Cu ft of vapor per lb at 60°F	8.66	6.51
m^3 of vapor per kg at 15.5°C	0.534	0.410
Specific gravity of vapor (Air = 1) at 60°F (15.5°C)	1.50	2.01
Ignition temperature in air	920–1120°F (493–549°C)	900–1000°F (482–538°C)
Maximum flame temperature in air	3595°F (1980°C)	3615°F (1990°C)
Limits of flammability in air, percent of vapor in air–gas mixture		
(a) Lower	2.15	1.55
(b) Upper	9.60	8.60
Latent heat of vaporization at boiling point		
(a) Btu/lb	184	167
(b) Btu/gal	773	808
(c) kJ/kg	428	388
(d) kJ/L	216	226
Total heating values after vaporization		
(a) Btu/cu ft	2,488	3,280
(b) Btu/lb	21,548	21,221
(c) Btu/gal	91,547	102,032
(d) kJ/m^3	92,430	121,280
(e) kJ/kg	49,920	49,140
(f) kJ/m^3	25,140	28,100

Source: *Fire Protection Handbook,* 19th edition, Table 8.7.5.

Properties of Materials Related to Motor Vehicle Fires

Tables 4.28 through 4.31 present selected data on materials associated with motor vehicle fires.

Properties of Materials

Table 4.28 Properties of Ignitable Liquids in Motor Vehicle Fires

Liquid	Flash Point		Ignition Temperature		Flammability Range (%)		Boiling Point		Vapor Density (Air = 1)
	°F	°C	°F	°C	Lower	Upper	°F	°C	
Brake fluid[a]	240–355	115–179							
Brake fluid[b]	298	148							
Ethylene glycol (100%)[c]	232	111	775	413	3.3	—	485	252	
Ethylene glycol (90%)[c]	270	132					387	197	
Diesel #2D[d]	126–204	52–96	494	257					
Kerosene #1 fuel oil[d]	100–162	38–72	410	210	0.7	5.0	304–574	151–301	
Gasoline—100 octane[d]	–36	–38	853	456	1.4	7.6	100–400	38–204	3–4
Methanol[d]	52	11	867	464	7.8	86.0	147	64	1.1
Motor oil[e]	410–495	210–257	500–700	260–371					
Trans fluid[e]	350	177							
Trans fluid[b]	361–379	183–193	410–417	210–214					
Dextron IIE									
Dextron II	367	186	414	212					
Type F (Ford)	347	175							
Power steering fluid[e]	350	177							

Note: The data provided in this table are for generic or typical products when tested in a specific way. The test methods may not be the same for each material. The information in this table is from various sources within published literature.

[a]NFPA SPP 51, Flash Point Index of Trade Name Liquids, p. 182.
[b]UNOCAL Lub Oils and Greases Div.
[c]Flick, Noyes Data Corp., Industrial Solvents Handbook, p. 416.
[d]NFPA Fire Protection Guide to Hazardous Materials.
[e]Severy, Blaisdell, and Kerkhoff, Automobile Collision Fires.
Source: NFPA 921, 2001 edition, Table 22.3.1.

Table 4.29 Gaseous Fuels in Motor Vehicles

Gas	Ignition Temperature °F	Ignition Temperature °C	Boiling Point °F	Boiling Point °C	Flammability Range % Lower	Flammability Range % Upper	Vapor Density (Air = 1)
Hydrogen	932	500	−422	−252	4.0	75.0	0.1
Natural gas (methane)	999	537	−259	−162	5.0	15.0	0.6
Propane gas	842	450	−44	−42	2.1	9.5	1.6

Note: The data provided in this table are for generic or typical products and may not represent the values for a specific product. When possible, values specific to the product involved should be obtained from a material safety data sheet or by test.
Source: NFPA 921, 2001 edition, Table 22.3.2.

Table 4.30 Solid Fuels in Motor Vehicle Fires

Material	Ignition Temperature °F	Ignition Temperature °C	Melting Point °F	Melting Point °C	Comments
Acrylic fibers	1040	560[b]	122	50[b]	
Aluminum (pure)	1832	1000[d*]	1220	660[d*]	
ABS	871	466[b]	230–257	110–125[c]	Body panels—may be completely consumed
Fiberglass (polyester resin)	1040	560[b]	802–932	428–500[d]	Resin burns but not glass body panels
Magnesium (pure)	1153	623[d*]	1202	650[c*]	
Nylon[a]	790	421[b]	349–509	176–265[c]	Trim, window gears, timing gears
Polyethylene	910	488[e*]	251–275	122–135[f]	Wiring insulation
Polystyrene	1063	573[e]	248–320	120–160[f]	Insulation, padding, trim
Polyurethane—foam	852–1074	456–579[e]			Seats, arm rests, padding
Polyurethane—rigid	590	310[b]	248–320	120–160[c]	Trim
Vinyl (PVC)	945	507[e]	167–221	75–105[f]	Wire insulation, upholstery

Note: The data provided in this table are for generic or typical products and may not represent the values for a specific product. When possible, values specific to the product involved should be obtained from the manufacturer or by test.
*Pure metal
[a]Lide (ed.), *Handbook of Chemistry and Physics.*
[b]Hilado, *Flammability Handbook for Plastics.*
[c]*Guide to Plastics.*
[d]NFPA *Fire Protection Handbook,* Table 3.13A (17th edition).
[e]NFPA *Fire Protection Handbook,* Table A.6 (17th edition).
[f]*Plastics Handbook.*
Source: NFPA 921, 2001 edition, Table 22.3.3.

INFORMATION FOR THE FIRE INVESTIGATOR
Properties of Materials

INFORMATION FOR THE FIRE INVESTIGATOR
Properties of Materials

Table 4.31 Flammable and Combustible Liquids Commonly Found in Service Stations

Liquid	Flash Point	NFPA 30 Class	Boiling Point	Minimum Ignition Temperature in Air
Antifreeze	230°F (110°C)	IIIB	300°F (149°C)	—
Brake fluid	300°F (149°C)	IIIB	540°F (282°C)	—
Chassis grease	400°F (204°C)	IIIB	>800°F (427°C)	>800°F (427°C)
Crankcase drainings	—	IIIB	—	—
Diesel fuel no. 1	100°F (38°C)	II	—	—
Diesel fuel no. 2	125°F (52°C)	II	—	—
Diesel fuel no. 4	130°F (54°C)	II	—	—
Gasoline	−40 to −50°F (−40 to −46°C)	IB	100 to 400°F (38 to 204°C)	~825°F (441°C)
Gear lubricant	395°F (202°C)	IIIB	>800°F (427°C)	>800°F (427°C)
Kerosene (fuel oil no. 1)	100°F (38°C)	II	303 to 574°F (151 to 301°C)	440°F (227°C)
Lithium-moly grease	380°F (193°C)	IIIB	>800°F (427°C)	>900°F (482°C)
Lubricating oils	300 to 450°F (149 to 232°C)	IIIB	—	—
Power steering fluid	350°F (177°C)	IIIB	>550°F (288°C)	—
Transmission fluid				
Dexon II	395°F (202°C)	IIIB	>800°F (427°C)	>822°F (427°C)
Type F	380°F (193°C)	IIIB	>800°F (427°C)	>822°F (427°C)
White grease	465°F (241°C)	IIIB	>800°F (427°C)	>822°F (427°C)
Windshield Washer Fluid (methanol/water mixtures)				
100% methanol	54°F (12°C)	IB	148°F (64°C)	725°F (385°C)
50% methanol	80°F (27°C)	IB	—	—
20% methanol	118°F (48°C)	II	—	—
5% methanol	206°F (97°C)	IIIB	—	—

Source: *Flammable and Combustible Liquids Code Handbook,* 6th edition, Table 12-1.

BURN INJURY AND OVERPRESSURE DAMAGE

Documenting the location and degree of fire victim burn or overpressure injuries can be useful in evaluating origin and cause for fires and explosions. The pressures associated with the degree of structural damage can also be helpful in analysis of explosion incidents. Tables 4.32 through 4.34 document burn injuries and present data to associate the level of injury or property damage with levels of overpressure. Background information and references on interpretation of human injury and overpressure damage are presented in Chapter 20 of NFPA 921.

Tables 4.34 and 4.35, respectively, present criteria for property damage and for high explosives overpressure.

Table 4.32 Percentage of Body Surface Area

Body Part	Infant	Child	Adult
Front of head	9.5	8.5	3.5
Rear of head	9.5	8.5	3.5
Front of neck	1.0	1.0	1.0
Rear of neck	1.0	1.0	1.0
Chest and abdomen	13.0	13.0	13.0
Genitalia	1.0	1.0	1.0
Back and buttocks	17.0	17.0	17.0
Front of arm and hand	4.25	4.25	4.75
Rear of arm and hand	4.25	4.25	4.75
Front of leg and foot	6.25	6.75	10.0
Rear of leg and foot	6.25	6.75	10.0

Note: Infant = up to age 4; child = age 5 to 10; adult = age 11 and above.
Source: NFPA 921, 2001 edition, Table 20.7.2.2.

Table 4.33 Human Injury Criteria

Overpressure (psi)	Injury	Comments	Source
0.6	Threshold for injury from flying glass*	Based on studies using sheep and dogs	a
1.0–2.0	Threshold for skin laceration from flying glass	Based on U.S. Army data	b
1.5	Threshold for multiple skin penetrations from flying glass (bare skin)*	Based on studies using sheep and dogs	a
2.0–3.0	Threshold for serious wounds from flying glass	Based on U.S. Army data	b
2.4	Threshold for eardrum rupture	Conflicting data on eardrum rupture	b
2.8	10% probability of eardrum rupture	Conflicting data on eardrum rupture	b
3.0	Overpressure will hurl a person to the ground	One source suggested an overpressure of 1.0 psi for this effect	c
3.4	1% eardrum rupture	Not a serious lesion	d
4.0–5.0	Serious wounds from flying glass near 50% probability	Based on U.S. Army data	b
5.8	Threshold for body-wall penetration from flying glass (bare skin)*	Based on studies using sheep and dogs	a
6.3	50% probability of eardrum rupture	Conflicting data on eardrum rupture	b
7.0–8.0	Serious wounds from flying glass near 100% probability	Based on U.S. Army data	b
10.0	Threshold lung hemorrhage	Not a serious lesion [applies to a blast of long duration (over 50 m/s)]; 20–30 psi required for 3 m/s duration waves	d
14.5	Fatality threshold for direct blast effects	Fatality primarily from lung hemorrhage	b
16.0	50% eardrum rupture	Some of the ear injuries would be severe	d
17.5	10% probability of fatality from direct blast effects	Conflicting data on mortality	b
20.5	50% probability of fatality from direct blast effects	Conflicting data on mortality	b
25.5	90% probability of fatality from direct blast effects	Conflicting data on mortality	b
27.0	1% mortality	A high incidence of severe lung injuries [applies to a blast of long duration (over 50 m/s)]; 60–70 psi required for 3 m/s duration waves	d
29.0	99% probability of fatality from direct blast effects	Conflicting data on mortality	b

Note: For SI units, 1 psi = 6.9 kPa.
*Interpretation of tables of data presented in reference.
[a]Fletcher, Richmond, and Yelverron, 1980.
Source: NFPA 921, 2001 edition, Table 18.13.3.1(a).

[b]Loss Prevention in the Process Industries.
[c]Brasie and Simpson, 1968.
[d]U.S. Department of Transportation, 1988.

Table 4.34 Property Damage Criteria

Overpressure (psi)	Damage	Source
0.03	Occasional breaking of large glass windows already under strain	a
0.04	Loud noise (143 dB). Sonic boom glass failure	a
0.10	Breakage of small windows, under strain	a
0.15	Typical pressure for glass failure	a
0.30	"Safe distance" (probability 0.95 no serious damage beyond this value) Missile limit Some damage to house ceilings 10% window glass broken	a
0.4	Minor structural damage	a, c
0.5–1.0	Shattering of glass windows, occasional damage to window frames. One source reported glass failure at 0.147 psi (1 kPa)	a, c, d, e
0.7	Minor damage to house structures	a
1.0	Partial demolition of houses, made uninhabitable	a
1.0–2.0	Shattering of corrugated asbestos siding Failure of corrugated aluminum-steel paneling Failure of wood siding panels (standard housing construction)	a, b, d, e
1.3	Steel frame of clad building slightly distorted	a
2.0	Partial collapse of walls and roofs of houses	a
2.0–3.0	Shattering of nonreinforced concrete or cinder block wall panels [1.5 psi (10.3 kPa) according to another source]	a, b, c, d
2.3	Lower limit of serious structural damage	a
2.5	50% destruction of brickwork of house	a
3.0	Steel frame building distorted and pulled away from foundations	a
3.0–4.10	Collapse of self-framing steel panel buildings Rupture of oil storage tanks Snapping failure—wooden utility tanks	a, b, c
4.0	Cladding of light industrial buildings ruptured	a
4.8	Failure of reinforced concrete structures	e
5.0	Snapping failure—wooden utility poles	a, b
5.0–7.0	Nearly complete destruction of houses	a
7.0	Loaded train wagons overturned	a
7.0–8.0	Shearing/flexure failure of brick wall panels [8 in. to 12 in. (20.3 cm to 30.5 cm) thick, not reinforced]	a, b, c, d
	Sides of steel frame buildings blown in	d
	Overturning of loaded rail cars	b, c
		Continued

INFORMATION FOR THE FIRE INVESTIGATOR
Burn injury and Overpressure Damage

INFORMATION FOR THE FIRE INVESTIGATOR
Burn injury and Overpressure Damage

Table 4.34 Property Damage Criteria *(continued)*

Overpressure (psi)	Damage	Source
9.0	Loaded train boxcars completely demolished	a
10.0	Probable total destruction of buildings	a
30.0	Steel towers blown down	b, c
88.0	Crater damage	e

[a]*Loss Prevention in the Process Industries.*
[b]Brasie and Simpson, 1968.
[c]U.S. Department of Transportation, 1988.
[d]U.S. Air Force, 1983.
[e]McRae, 1984.
Source: NFPA 921, 2001 edition, Table 18.13.3.1(b).

Table 4.35 High Explosives Overpressure Constants and Consequences

Scaled Distance Z (ft/kg$^{1/3}$)	Overpressure (psi)	Consequences
3000–890	0.01–0.04	Minimum damage to glass panels
420–200	0.1–0.2	Typical window glass breakage
200–100	0.2–0.4	Minimum overpressure for debris and missile damage
82–41	0.5–1.1	Windows shattered, plaster cracked, minor damage to some buildings
44–32	1.0–1.5	Personnel knocked down
44–28	1.0–1.8	Panels of sheet metal buckled
44–24	1.0–2.2	Failure of wooden siding for conventional homes
28–20	1.8–2.9	Failure of walls constructed of concrete blocks or cinder blocks
20–16	2.9–4.4	Self-framing paneled buildings collapse
20–16	2.9–4.4	Oil storage tanks ruptured
16–12	4.4–7.3	Utility poles broken off
16–12	4.4–7.3	Serious damage to buildings with structural steel framework
11–10	10.2–11.6	Probable total destruction of most buildings
15–9	5.1–14.5	Eardrum rupture
14–11	5.8–8.7	Reinforced concrete structures severely damaged
14–11	5.8–8.7	Railroad cars overturned
6.7–4.5	29.0–72.5	Lung damage
3.8–2.7	102–218	Lethality
2.4–1.9	290–435	Crater formation in average soil

Source: *Fire Protection Handbook,* 19th edition, Table 2.8.1, from Kinney, G. and Graham, K., *Explosive Shocks in Air,* 2nd edition, Springer-Verlag, 1985.

OTHER BASIC INFORMATION FOR THE FIRE INVESTIGATOR

Part 4 continues by presenting key data that cannot be categorized as solely a property of ignition or as one form of material. See Tables 4.36 through 4.41 and Figure 4.1.

Table 4.36 Physical and Combustion Properties of Selected Fuels in Air

Fuel	Mol. wt.	Spec. grav.	T_{Boil} (°C)	Heat of vap. (kJ/kg)	Heat of comb. (mJ/kg)	Stoichiometry		Flammability Limits (% stoichio.)	
						% Vol.	f^a	Lean	Rich
Acetaldehyde	44.1	0.783	–56.7	569.4	—	0.0772	0.1280	—	—
Acetone	58.1	0.792	56.7	523.0	30.8	0.0497	0.1054	59	233
Acetylene	26.0	0.621	–83.9	—	48.2	0.0772	0.0755	31	—
Acrolein	56.1	0.841	52.8	—	—	0.0564	0.1163	48	752
Acrylonitrile	53.1	0.797	78.3	—	—	0.0528	0.1028	87	—
Ammonia	17.0	0.817	–33.3	1373.6	—	0.2181	0.1645	—	—
Aniline	93.1	1.022	184.4	432.6	—	0.0263	0.0872	—	—
Benzene	78.1	0.885	80.0	431.8	39.9	0.0277	0.0755	43	336
Benzyl alcohol	108.1	1.050	205.0	—	—	0.0240	0.0923	—	—
1,2-Butadiene (methylallene)	54.1	0.658	11.1	—	45.5	0.0366	0.0714	—	—
n-Butane	58.1	0.584	–0.5	385.8	45.7	0.0312	0.0649	54	330
Butanone (methylethyl ketone)	72.1	0.805	79.4	—	—	0.0366	0.0951	—	—
1-Butene	56.1	0.601	–6.1	443.9	45.3	0.0377	0.0678	53	353
d-Camphor	152.2	0.990	203.4	—	—	0.0153	0.0818	—	—
Carbon disulfide	76.1	1.263	46.1	351.0	—	0.0652	0.1841	18	1120
Carbon monoxide	28.0	—	–190.0	211.7	—	0.2950	0.4064	34	676
Cyclobutane	56.1	0.703	12.8	—	—	0.0377	0.0678	—	—
Cyclohexane	84.2	0.783	80.6	258.1	43.8	0.0227	0.0678	48	401
Cyclohexene	82.1	0.810	82.8	—	—	0.0240	0.0701	—	—
Cyclopentane	70.1	0.751	49.4	388.3	44.2	0.0271	0.0678	—	—
Cyclopropane	42.1	0.720	–34.4	—	—	0.0444	0.0678	58	276
trans-Decalin	138.2	0.874	187.2	—	—	0.0142	0.0692	—	—
n-Decane	142.3	0.734	174.0	359.8	44.2	0.0133	0.0666	45	356
Diethyl ether	74.1	0.714	34.4	351.6	—	0.0337	0.0896	55	2640

Table continues to the right

Fuel	Spont. Ign. Temp. (°C)	Fuel for Max. Flame Speed (% stoichio.)	Max. Flame Speed (cm/s)	Flame Temp. at Max. Fl. Speed K	Ign. Energy Stoich. (10^{-5} cal.)	Ign. Energy Min.	Quenching Dist. Stoich. (mm)	Quenching Dist. Min.
Acetaldehyde	—	131	50.18	2121	8.99	—	2.29	—
Acetone	561.1	133	—	—	27.48	—	3.81	—
Acetylene	305.0	100	155.25	2461	0.72	—	0.76	—
Acrolein	277.8	105	61.75	2600	4.18	—	1.52	—
Acrylonitrile	481.1	—	46.75	—	8.60	3.82	2.29	1.52
Ammonia	651.1	—	—	—	—	—	—	—
Aniline	593.3	108	—	—	—	—	—	—
Benzene	591.7	—	44.60	2365	13.15	5.38	2.79	1.78
Benzyl alcohol	427.8	117	—	—	5.60	—	1.30	—
1,2-Butadiene (methylallene)	—	113	63.90	2419	—	—	—	—
n-Butane	430.6	100	41.60	2256	18.16	6.21	3.05	1.78
Butanone (methylethyl ketone)	—	116	39.45	—	12.67	6.69	2.54	2.03
1-Butene	443.3	—	47.60	2319	—	—	—	—
d-Camphor	466.1	102	—	—	—	—	—	—
Carbon disulfide	120.0	170	54.46	—	0.36	—	0.51	—
Carbon monoxide	608.9	—	42.88	—	—	—	—	—
Cyclobutane	—	115	62.18	2250	32.98	5.33	4.06	1.78
Cyclohexane	270.0	117	42.46	2308	20.55	—	3.30	—
Cyclohexene	—	—	44.17	—	—	—	—	—
Cyclopentane	385.0	117	41.17	2264	19.84	—	3.30	—
Cyclopropane	497.8	113	52.32	2328	5.74	5.50	1.78	1.78
trans-Decalin	271.7	109	33.88	2222	—	—	2.06	—
n-Decane	231.7	105	40.31	2286	—	—	—	—
Diethyl ether	185.6	115	43.74	2253	11.71	6.69	2.54	2.03

Continued

Table 4.36 Physical and Combustion Properties of Selected Fuels in Air (continued)

Fuel	Mol. wt.	Spec. grav.	T_{Boil} (°C)	Heat of vap. (kJ/kg)	Heat of comb. (mJ/kg)	Stoichiometry		Flammability Limits (% stoichio.)	
						% Vol.	f^a	Lean	Rich
Ethane	30.1	—	−88.9	488.3	47.4	0.0564	0.0624	50	272
Ethyl acetate	88.1	0.901	77.2	—	—	0.0402	0.1279	61	236
Ethanol	46.1	0.789	78.5	836.8	26.8	0.0652	0.1115	—	—
Ethylamine	45.1	0.706	16.7	611.3	—	0.0528	0.0873	—	—
Ethylene oxide	44.1	1.965	10.6	581.1	—	0.0772	0.1280	—	—
Furan	68.1	0.936	32.2	400.0	—	0.0444	0.1098	—	—
n-Heptane	100.2	0.688	98.5	364.9	44.4	0.0187	0.0661	53	450
n-Hexane	86.2	0.664	68.0	364.9	44.7	0.0216	0.0659	51	400
Hydrogen	2.0	—	−252.7	451.0	119.9	0.2950	0.0290	—	—
iso-Propanol	60.1	0.785	82.2	664.8	—	0.0444	0.0969	—	—
Kerosene	154.0	0.825	250.0	290.8	43.1	—	—	—	—
Methane	16.0	—	−161.7	509.2	50.0	0.0947	0.0581	46	164
Methanol	32.0	0.793	64.5	1100.9	19.8	0.1224	0.1548	48	408
Methyl formate	60.1	0.975	31.7	472.0	—	0.0947	0.2181	—	—
n-Nonane	128.3	0.772	150.6	288.3	44.6	0.0147	0.0665	47	434
n-Octane	114.2	0.707	125.6	300.0	44.8	0.0165	0.0633	51	425
n-Pentane	72.1	0.631	36.0	364.4	45.3	0.0255	0.0654	54	359
1-Pentene	70.1	0.646	30.0	—	45.0	0.0271	0.0678	47	370
Propane	44.1	0.508	−42.2	425.5	46.3	0.0402	0.0640	51	283
Propene	42.1	0.522	−47.7	437.2	45.8	0.0444	0.0678	48	272
n-Propanol	60.1	0.804	97.2	685.8	—	0.0444	0.0969	—	—
Toulene	92.1	0.872	110.6	362.8	40.9	0.0227	0.0743	43	322
Triethylamine	101.2	0.723	89.4	—	—	0.0210	0.0753	—	—
Turpentine	—	—	—	—	—	—	—	—	—

Table continues to the right ⬅

Fuel	Spont. Ign. Temp. (°C)	Fuel for Max. Flame Speed (% stoichio.)	Max. Flame Speed (cm/s)	Flame Temp. at Max. Fl. Speed K	Ign. Energy (10⁻⁵ cal.)		Quenching Dist. (mm)	
					Stoich.	Min.	Stoich.	Min.
Ethane	472.2	112	44.17	2244	10.04	5.74	2.29	1.78
Ethyl acetate	486.1	100	35.59	—	33.94	11.47	4.32	2.54
Ethanol	392.2	—	—	—	—	—	—	—
Ethylamine	—	—	—	—	57.36	—	5.33	—
Ethylene oxide	428.9	125	11.35	2411	2.51	1.48	1.27	1.02
Furan	—	—	—	—	5.40	—	1.78	—
n-Heptane	247.2	122	42.46	2214	27.49	5.74	3.81	1.78
n-Hexane	260.6	117	42.46	2239	22.71	5.50	3.56	1.78
Hydrogen	571.1	170	291.19	2380	0.36	0.36	0.51	0.51
iso-Propanol	455.6	100	38.16	—	15.54	—	2.79	—
Kerosene	—	—	—	2236	—	—	—	—
Methane	632.2	106	37.31	2236	7.89	6.93	2.54	2.03
Methanol	470.0	101	52.32	—	5.14	3.35	1.78	1.52
Methyl formate	—	—	—	—	14.82	—	2.79	—
n-Nonane	238.9	—	—	2251	—	—	—	—
n-Octane	240.0	—	—	—	—	—	—	—
n-Pentane	284.4	115	42.46	2250	19.60	5.26	3.30	1.78
1-Pentene	298.3	114	46.75	2314	—	—	—	—
Propane	504.4	114	42.89	2250	7.29	—	2.03	1.78
Propene	557.8	114	48.03	2339	6.74	—	2.03	—
n-Propanol	433.3	—	—	—	—	—	—	—
Toluene	567.8	—	38.60	2344	—	—	—	—
Triethylamine	—	105	—	—	27.48	—	3.81	—
Turpentine	252.2	—	—	—	—	—	—	—

Continued

Table 4.36 Physical and Combustion Properties of Selected Fuels in Air (continued)

Fuel	Mol. wt.	Spec. grav.	T_{Boil} (°C)	Heat of vap. (kJ/kg)	Heat of comb. (mJ/kg)	Stoichiometry % Vol.	Stoichiometry f^a	Flammability Limits (% stoichio.) Lean	Flammability Limits (% stoichio.) Rich
Xylene	106.0	0.870	130.0	334.7	43.1	—	—	—	—
Gasoline 73 octane	120.0	0.720	155.0	338.9	44.1	—	—	—	—
Gasoline 100 octane	—	—	—	—	—	—	—	—	—
Jet fuel JP1	150.0	0.810	—	—	43.0	0.0130	0.0680	—	—
JP3	112.0	0.760	—	—	43.5	0.0170	0.0680	—	—
JP4	126.0	0.780	—	—	43.5	0.0150	0.0680	—	—
JP5	170.0	0.830	—	—	43.0	0.0110	0.0690	—	—

Table continues to the right

$^a f$ is the stoichiometric air/fuel ratio; i.e., $f = 1/r$.

Source: Table C.1, *SFPE Handbook of Fire Protection Engineering*, 3rd edition, Courtesy of the Society of Fire Protection Engineers, Bethesda, MD.

Fuel	Spont. Ign. Temp. (°C)	Fuel for Max. Flame Speed (% stoichio.)	Max. Flame Speed (cm/s)	Flame Temp. at Max. Fl. Speed K	Ign. Energy (10^{-5} cal.)		Quenching Dist. (mm)	
					Stoich.	Min.	Stoich.	Min.
Xylene	298.9	—	—	—	—	—	—	—
Gasoline 73 octane	468.3	106	37.74	—	—	—	—	—
Gasoline 100 octane	248.9	107	36.88	—	—	—	—	—
Jet fuel JP1	—	107	—	—	—	—	—	—
JP3	—	—	—	—	—	—	—	—
JP4	261.1	107	38.17	—	—	—	—	—
JP5	242.2	—	—	—	—	—	—	—

INFORMATION FOR THE FIRE INVESTIGATOR
Other Basic Information for the Fire Investigator

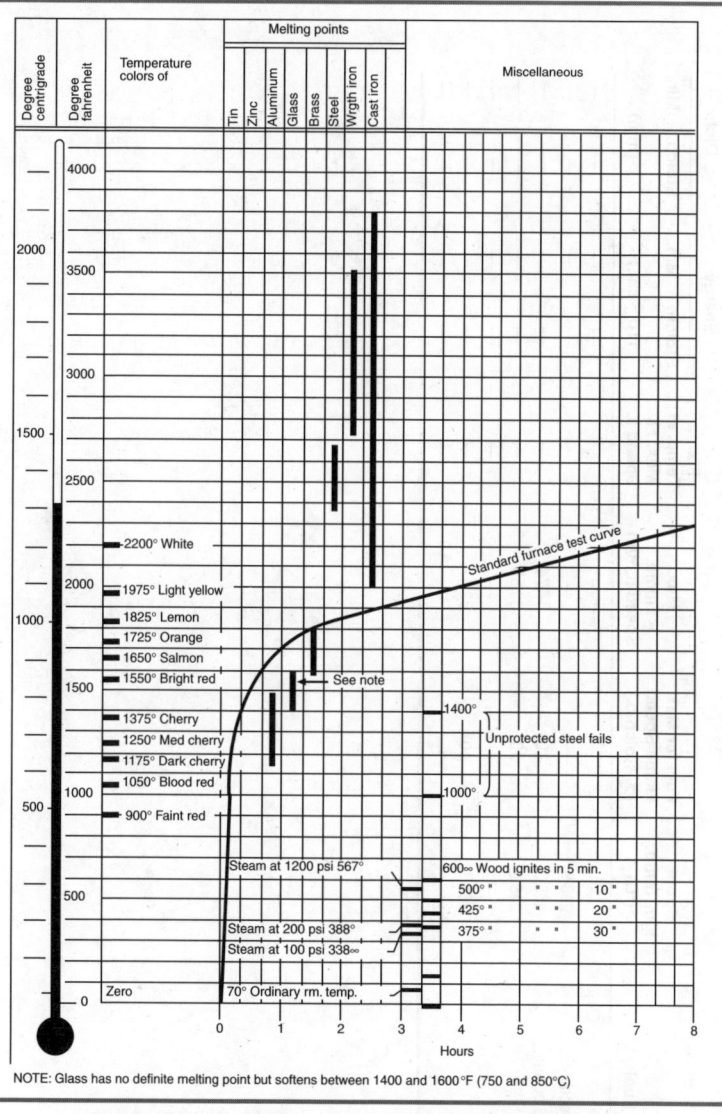

NOTE: Glass has no definite melting point but softens between 1400 and 1600 °F (750 and 850°C)

Figure 4.1 Temperature colors, melting points of various substances, and data on the behavior of some materials at elevated temperatures and pressures.
Source: *Fire Protection Handbook,* 19th edition, Figure A.1.

Table 4.37 Representative Peak Heat Release Rates (unconfined burning)

Fuel (lb)	Peak HRR (kW)
Wastebasket, small (1.5–3)	4–18
Trash bags, 11 gal with mixed plastic and paper trash (2-1/2–7-1/2)	140–350
Cotton mattress (26–29)	40–970
TV sets (69–72)	120–290
Plastic trash bags/paper trash (2.6–31)	120–350
PVC waiting room chair, metal frame (34)	270
Cotton easy chair (39–70)	290–370
Gasoline/kerosene in 2 ft^2 (0.61 m^2) pool	400
Christmas trees, dry (14–16)	500–650
Polyurethane mattress (7–31)	810–2630
Polyurethane easy chair (27–61)	1350–1990
Polyurethane sofa (113)	3120

Values are from the following publications:

Babrauskas and Krasny, *Fire Behavior of Upholstered Furniture.*

NFPA 72, National Fire Alarm Code®, 1996 ed., B.2.2.2.1.

Lee, *Heat Release Rate Characteristics of Some Combustible Fuel Sources in Nuclear Power Plants.*

Source: NFPA 921, 2001 edition, Table 3.4.

Table 4.38 Approximate Melting Temperatures of Common Materials

Material	°F	°C	Material	°F	°C
Aluminum (alloys)[b]	1050–1200	566–650	Plastics (thermo) *(cont'd)*		
Aluminum[a]	1220	660	Nylon[d]	349–509	176–265
Brass (yellow)[b]	1710	932	Polyethylene[d]	251–275	122–135
Brass (red)[b]	1825	996	Polystyrene[d]	248–320	120–160
Bronze (aluminum)[b]	1800	982	Polyvinylchloride[d]	167–221	75–105
Cast iron (gray)[a]	2460–2550	1350–1400	Platinum[a]	3224	1773
Cast iron (white)[a]	1920–2010	1050–1100	Porcelain[a]	2820	1550
Chromium[a]	3350	1845	Pot metal[e]	562–752	300–400
Copper[a]	1981	1082	Quartz (SiO$_2$)[a]	3060–3090	1682–1700
Fire brick (insulating)[a]	2980–3000	1638–1650	Silver[a]	1760	960
Glass[a]	1100–2600	593–1427	Solder (tin)[a]	275–350	135–177
Gold[a]	1945	1063	Steel (stainless)[b]	2600	1427
Iron[a]	2802	1540	Steel (carbon)[b]	2760	1516
Lead[a]	621	327	Tin[a]	449	232
Magnesium (AZ31B alloy)[b]	1160	627	Wax (paraffin)[c]	120–167	49–75
Nickel[a]	2651	1455	White pot metal[e]	562–752	300–400
Paraffin[a]	129	54	Zinc[a]	707	375
Plastics (thermo)					
ABS[d]	190–257	88–125			
Acrylic[d]	194–221	90–105			

[a]Baumeister, Avallone, and Baumeister III, *Mark's Standard Handbook for Mechanical Engineers.*

[b]Lide, ed., *Handbook of Chemistry and Physics.*

[c]NFPA *Fire Protection Guide to Hazardous Materials.*

[d]*Plastics Handbook.*

[e]Glick and Gieck, *Engineering Formulas.*

Source: NFPA 921, 2001 edition, Table 4.8.

Table 4.39 Approximate Rate of Radiant Flux

Approximate Radiant Heat Flux (kW/m^2)	Comment or Observed Effect
170	Maximum heat flux as currently measured in a postflashover fire compartment.
80	Heat flux for protective clothing Thermal Protective Performance (TPP) Test.[a]
52	Fiberboard ignites spontaneously after 5 seconds.[b]
29	Wood ignites spontaneously after prolonged exposure.[b]
20	Heat flux on a residential family room floor at the beginning of flashover.[c]
16	Human skin experiences sudden pain and blisters after 5-second exposure with second-degree burn injury.[a]
12.5	Wood volatiles ignite with intended exposure[d] and piloted ignition.
10.4	Human skin experiences pain with 3-second exposure and blisters in 9 seconds with second-degree burn injury.[a,b]
6.4	Human skin experiences pain with a second exposure and blisters in 18 seconds with second-degree burn injury.[a,e]
4.5	Human skin becomes blistered with a 30-second exposure, causing a second-degree burn injury.[a]
2.5	Common thermal radiation exposure while fire fighting.[f] This energy level may cause burn injuries with prolonged exposure.
≤1.4	Thermal radiation from the sun. Potential sunburn in 30 minutes or less.[g]

Note: The unit kW/m^2 defines the amount of heat energy or flux that strikes a known surface area of an object. The unit (kW) represents 1000 watts of energy and the unit (m^2) represents the surface area of a square measuring 1 m long and 1 m wide. For example, 1.4 kW/m^2 represents 1.4 multiplied by 1000 and equals 1400 watts of energy. This surface area may be that of the human skin or any other material.

[a]NFPA 1971, *Standard on Protective Ensemble for Structural Fire Fighting.*

[b]Lawson, "Fire and the Atomic Bomb."

[c]Fang and Breese, "Fire Development in Residential Basement Rooms."

[d]Lawson and Simms, "The Ignition of Wood by Radiation," pp. 288–292.

[e]Tan, "Flare System Design Simplified," pp. 172–176.

[f]U.S. Fire Administration, "Minimum Standards on Structural Fire Fighting Protective Clothing and Equipment."

[g]Bennet and Myers, "Momentum, Heat, and Mass Transfer."

Source: NFPA 921, 2001 edition, Table 3.5.3.2.

Table 4.40 Miscellaneous Properties of Plastics and Other Combustibles

Material	Density (kg/m³)	Min. Radiant Flux for Ignition (kW/m²)	Min. Mass Loss Rate for Ignition (g/m²-s)	Energy Required for Ignition (kJ/m²)	Ignition Temperature (K)	Effective Heat of Gasification (MJ/kg)	Smoke Specific Extinction Area (m²/kg)	Soot Yield (kg soot/kg specimen)
ABS						3.2		
benzene, liquid							830–900	
cellulose						3.5	10–15	
Douglas Fir						1.8		
fiberglass, FR	—	7	—	650–770		1.7–2.2	250–370	
nylon 6/6				3950	740	2.3	410	0.14
paper, corrugated						2.2		
phenolic, foams, rigid	67	45	5.5	390	944	3.7		
polycarbonate						2.1		
polyethylene	920	19	2.5	1500–5100	761	1.5–2.7	290–510	
polyethylene, foams	36–143	19–22	2.5–2.6	1430–1780	761–807	1.5–2.7	360–810	
polyethylene, foam, chlorinated						2.1–3.1	1010–1360	
polyisocyanurate, foams, rigid	33–36	23	5.5–6.8	850–930	798	4.5		
polymethylmethacrylate	1170	18	4.4	1300–4200	751	1.6	110–260	0.012
polyoxymethylene	1400	17	4.5	2100–6000	740	2.4	0–10	
polypropylene	905	20	2.7	1000–3500	771	1.4–2.0	380–610	0.07–0.09
polystyrene	1050	29	4.0	1300–6400	846	1.7	730–1430	
polystyrene, foams, rigid	16–34	18–27	4.9–6.3	1200–3200	751–831	1.3–3.1	940–1300	
polytetrafluoroethylene		43		9000	933			0.02
polyurethane, foams, flexible	29–47	16–30	5.3–7.2	150–770	729–852	1.2–2.7	100–500	

Continued

Table 4.40 Miscellaneous Properties of Plastics and Other Combustibles (continued)

Material	Density (kg/m³)	Min. Radiant Ignition Flux for Ignition (kW/m²)	Min. Mass Loss Rate for Ignition (g/m²-s)	Energy Required for Ignition (kJ/m²)	Ignition Temperature (K)	Effective Heat of Gasification (MJ/kg)	Smoke Specific Extinction Area (m²/kg)	Soot Yield (kg soot/kg specimen)
polyurethane, foams, rigid	36–321	20–28	6.9–8.4	150–1300	771–838	2.0–4.5	100–1000	
polyvinyl chloride	1035	21	—	3320	780	1.7–2.5	1100–1400	0.23
red oak	740	11	2.5	420–920	730	1.7–5.5	80–130	.0002–0.018
styrene, liquid						0.64	780–920	

Source: *Fire Protection Handbook*, 19th edition, Table A.6.

Table 4.41 Ignition and Flammability Properties of Combustible Liquids and Gases in Air and Oxygen at Atmospheric Pressure

Combustible	Flashpoint Air °C (°F)		Min. Ign. Temperature		Min. Ign. Energy		Flammability Limits Vol. %			
		Air °C (°F)	Oxygen °C (°F)	Air mJ	Oxygen mJ	Air		Oxygen	
						LFL	UFL	LFL	UFL
Hydrocarbon Fuels										
Methane	Gas	630 (1166)	—	0.30	0.003	5.0	15	5.1	61
Ethane	Gas	515 (959)	506 (943)	0.25	0.002	3.0	12.4	3.0	66
n-Butane	-60 (-76)	288 (550)	278 (532)	0.25	0.009	1.8	8.4	1.8	49
n-Hexane	-3.9 (25)	225 (437)	218 (424)	0.288	0.006	1.2	7.4	1.2	52[a]
n-Octane	13.3 (56)	220 (428)	208 (406)	—	—	0.8	6.5	≤0.8	—
Ethylene	Gas	490 (914)	485 (905)	0.07	0.001	2.7	36	2.9	80
Propylene	Gas	458 (856)	423 (793)	0.28	—	2.4	11	2.1	53
Acetylene	Gas	305 (581)	296 (565)	0.017	0.0002	2.5	100	≤2.5	100
Gasoline (100/130)	-45.5 (-50)	440 (824)	316 (600)	—	—	1.3	7.1	≤1.3	—
Kerosene	37.8 (100)	227 (440)	216 (420)	—	—	0.7	5	0.7	—
Anesthetic Agents										
Cyclopropane	Gas	500 (932)	454 (849)	0.18	0.001	2.4	10.4	2.5	60
Ethyl ether	-28.9 (-20)	193 (380)	182 (360)	0.20	0.0013	1.9	36	2.0	82
Vinyl ether	-30 (<-22)	360 (680)	166 (331)	—	—	1.7	27	1.8	85
Ethylene	Gas	490 (914)	485 (905)	0.07	0.001	2.7	36	2.9	80
Ethyl chloride	-50 (-58)	516 (961)	468 (874)	—	—	4.0	14.8	4.0	67
Chloroform	—Nonflammable—									

Continued

Table 4.41 Ignition and Flammability Properties of Combustible Liquids and Gases in Air and Oxygen at Atmospheric Pressure (continued)

Combustible	Flashpoint Air °C (°F)	Min. Ign. Temperature Air °C (°F)	Min. Ign. Temperature Oxygen °C (°F)	Min. Ign. Energy Air mJ	Min. Ign. Energy Oxygen mJ	Flammability Limits Vol. % Air LFL	Air UFL	Oxygen LFL	Oxygen UFL
Anesthetic Agents (continued)									
Enflurane	>200 (93)	NA	NA	NA	NA	NA	NA	9.8	NA
Isoflurane	>200 (93)	NA	NA	NA	NA	NA	NA	8.8	NF
Desflurane	NF	NA	NA	NA	NA	NA	NA	NA	NA
Nitrous oxide	—Nonflammable—	NA	NA	NA	NA	NA	NA	17.2	20.8
Solvents									
Methyl alcohol	12.2 (54)	385 (725)	—	0.14	—	6.7	36	≤6.7	93
Ethyl alcohol	12.8 (55)	365 (689)	—	—	—	3.3	19	≤3.3	—
n-Propyl Alcohol	15 (59)	440 (824)	328 (622)	—	—	2.2	14	≤2.2	—
Glycol	111 (232)	400 (752)	—	—	—	3.5^a	—	≤3.5	—
Glycerol	160 (320)	370 (698)	320 (608)	0.48	—	—	—	—	—
Ethyl acetate	-4.4 (24)	427 (800)	—	—	—	2.2	11	≤2.2	—
n-Amyl acetate	24.4 (76)	360 (680)	234 (453)	—	—	1.0	7.1	≤1.0	—
Acetone	-17.8 (0)	465 (869)	—	1.15	0.0024	2.6	13	≤2.6	60^a
Benzene	-11.1 (12)	560 (1040)	—	0.22	—	1.3	7.9	≤1.3	30
Naphtha (Stoddard)	37.8 (~100)	232 (~450)	216 (~420)	2.5	—	1.0	6	≤1.0	—
Toluene	4.4 (40)	480 (896)	—	—	—	1.2	7.1	≤1.2	—
Butyl chloride	-6.7 (20)	240 (464)	235 (455)	0.332	0.007^a	1.8	10	1.7	52^a
Methylene chloride	—	615 (1139)	606 (1123)	—	0.137	15.9^a	19.1^a	11.7^a	68

Continued

Table 4.41 Ignition and Flammability Properties of Combustible Liquids and Gases in Air and Oxygen at Atmospheric Pressure (continued)

Combustible	Flashpoint Air °C (°F)	Min. Ign. Temperature Air °C (°F)	Oxygen °C (°F)	Min. Ign. Energy Air mJ	Oxygen mJ	Flammability Limits Vol. % Air LFL	UFL	Oxygen LFL	UFL
		Solvents (continued)							
Ethylene chloride	13.3 (56)	476 (889)	470 (878)	2.37	0.011[a]	6.2	16	4.0	67.5
Trichloroethane	40 (104)	458 (856)	418 (784)	—	0.092	6.3[a]	13[a]	5.5[a]	57[a]
Trichloroethylene	32.2 (90)	420 (788)	396 (745)	—	18[a]	10.5[a]	41[a]	7.5	91[a]
Carbon tetrachloride	—Nonflammable—								
		Miscellaneous Combustibles							
Acetaldehyde	−27.2 (−17)	175 (347)	159 (318)	0.38	—	4.0	60	4.0	93
Acetic Acid	40 (104)	465 (869)	—	—	—	5.4[a]	—	≤5.4	—
Ammonia	Gas	651 (1204)	—	>1000	—	15.0	28	15.0	79
Aniline	75.6 (168)	615 (1139)	—	—	—	1.2[a]	8.3	≤1.2	—
Carbon monoxide	Gas	609 (1128)	588 (1090)	—	—	12.5	74	≤12.5	94
Carbon disulfide	−30 (−22)	90 (194)	—	0.015	—	1.3	50	≤1.3	—
Ethylene oxide	<17.8 (<0)	429 (804)	—	0.062	—	3.6	100	≤3.6	100
Propylene oxide	−37.2 (−35)	—	400	0.14	—	2.8	37	≤2.8	—
Hydrogen	Gas	520 (968)	400 (752)	0.017	0.0012	4.0	75	4.0	95
Hydrogen sulfide	Gas	260 (500)	220 (428)	0.077	—	4.0	44	≤4.0	—
Bromochloromethane	—	450 (842)	368 (694)	—	—	NF[b]	NF	10.0	85

Continued

Table 4.41 Ignition and Flammability Properties of Combustible
Liquids and Gases in Air and Oxygen at Atmospheric Pressure (continued)

Combustible	Flashpoint Air °C (°F)	Min. Ign. Temperature Air °C (°F)	Min. Ign. Temperature Oxygen °C (°F)	Min. Ign. Energy Air mJ	Min. Ign. Energy Oxygen mJ	Flammability Limits Vol.% Air LFL	Air UFL	Oxygen LFL	Oxygen UFL
Miscellaneous Combustibles (continued)									
Bromotrifluoromethane	Gas	>593 (>1100)	657 (1215)	—	—	NF	NF	NF	NF
Dibromodifluoromethane	Gas	499 (930)	453 (847)	—	—	NF	NF	29.0	80

[a] Data at 200°F (93°C).

[b] NF—No flammable mixtures found in Reference 4.

Source: Fire Protection Handbook, 19th edition, Table 8.9.2.

SELECTED METRIC INFORMATION
FOR THE FIRE INVESTIGATOR

Table 4.42 displays English to SI units and SI to English units for units of length, area, mass and weight, liquid volume, temperature, velocity and speed, and energy that are most frequently used by the fire investigator.

Table 4.42 Selected SI Conversions for the Fire Investigator

To Convert from	To	Conversion Factor
Length		
Inches	Millimeters	× 25.4
Feet	Centimeters	× 30.48
Yards	Meters	× 0.9144
Miles	Kilometers	× 1.609
Millimeters	Inches	× 0.03937
Centimeters	Inches	× 0.3937
Meters	Yards	× 1.0936
Kilometers	Miles	× 0.6215
Area		
Square inches	Square centimeters	× 6.4452
Square feet	Square meters	× 0.09
Square yards	Square meters	× 0.8361
Square miles	Square kilometers	× 2.590
Acres	Hectares	× 0.4
Square centimeters	Square inches	× 0.155
Square meters	Square yards	× 1.196
Square kilometers	Square miles	× 0.3861
Hectares	Acres	× 2.5
Mass and Weight		
Fluid ounces	Grams	× 30
Pounds	Kilograms	× 0.4536
Tons	Metric tons	× 0.9
Grams	Fluid ounces	× 0.0333
Kilograms	Pounds	× 2.2046
Metric tons	Tons	× 1.1

Continued

Table 4.42 Selected SI Conversions for the Fire Investigator *(continued)*

To Convert from	To	Conversion Factor
Liquid Volume		
Ounces	Milliliters	× 30
Pints	Liters	× 0.568
Quarts	Liters	× 0.95
U.S. Gallons	Cubic meters	× 0.00378
U.S. Gallons	Liters	× 3.7854
Temperature		
Celsius (degrees C)	Fahrenheit (degrees F)	$9/5 \times C + 32$
Celsius (degrees C)	Kelvin (K)	$C + 273.15$
Fahrenheit (degrees F)	Celsius (degrees C)	$5/9 \times (t-32)$
Fahrenheit (degrees F)	Kelvin (K)	$5/9 \times (t + 459.67)$
Kelvin (K)	Celsius (degrees C)	$K - 273.15$
Kelvin (K)	Fahrenheit (degrees F)	$9/5 \times K - 459.67$
Velocity and Speed		
Feet per minute	Meters per second	× 0.00508
Kilometers per hour	Miles per hour	× 0.6214
Miles per hour	Meters per second	× 0.4470
Miles per hour	Kilometers per hour	× 1.6093
Energy		
British thermal unit/second (heat release rate)	Kilowatt (heat release rate)	× 1.05
Kilowatt (heat release rate)	British thermal unit/second (heat release rate)	× 0.95
Milliwatts	Watts	× 1000
Watts	Kilowatts	× 1000
Kilowatts	Megawatts	× 1000

APPENDIX

CHAPTER ORGANIZATION OF NFPA 921

All chapter, figure, and table references in the *Field Guide for Fire Investigators* are based on the 2001 edition of NFPA 921, *Guide for Fire and Explosion Investigations*. The content of NFPA 921 was re-organized for the 2004 edition to conform with the *NFPA Manual of Style*. To aid the *Field Guide* user, Table A.1 compares the table of contents of the 2001 and 2004 editions of NFPA 921 by chapter number. Table A.2 presents the same information, sorted alphabetically by chapter title.

Table A.1 Comparison of 2001 and 2004 Editions of NFPA 921, by Chapter Number

	2001 Edition	2004 Edition
Chapter 1	Administration	Administration
Chapter 2	Basic Methodology	Mandatory References
Chapter 3	Basic Fire Science	Definitions
Chapter 4	Fire Patterns	Basic Methodology
Chapter 5	Building Systems	Basic Fire Science
Chapter 6	Electricity and Fire	Fire Patterns
Chapter 7	Building Fuel Gas Systems	Building Systems
Chapter 8	Fire-Related Human Behavior	Electricity and Fire
Chapter 9	Legal Considerations	Building Fuel Gas Systems
Chapter 10	Safety	Fire-Related Human Behavior
Chapter 11	Sources of Information	Legal Considerations
Chapter 12	Planning the Investigation	Safety
Chapter 13	Recording the Scene	Sources of Information
Chapter 14	Physical Evidence	Planning the Investigation
Chapter 15	Origin Determination	Documenting the Investigation
Chapter 16	Cause Determination	Physical Evidence
Chapter 17	Failure Analysis and Analytical Tools	Origin Determination
Chapter 18	Explosions	Cause Determination

Continued

CHAPTER ORGANIZATION OF NFPA 921

Table A.1 Comparison of 2001 and 2004 Editions of NFPA 921, by Chapter Number *(continued)*

	2001 Edition	2004 Edition
Chapter 19	Incendiary Fires	Analyzing the Incident for Cause and Responsibility
Chapter 20	Fire and Explosion Deaths and Injuries	Failure Analysis and Analytical Tools
Chapter 21	Appliances	Explosions
Chapter 22	Motor Vehicle Fires	Incendiary Fires
Chapter 23	Wildfire Investigations	Fire and Explosion Deaths and Injuries
Chapter 24	Management of Major Investigations	Appliances
Chapter 25	Referenced Publications	Motor Vehicle Fires
Chapter 26	N/A	Wildfire Investigations
Chapter 27	N/A	Management of Major Investigations
Annex A	Appendix A	Explanatory Material
Annex B	Appendix B	Bibliography
Annex C	Appendix C	Referenced Publications

Table A.2 Comparison of 2001 and 2004 Editions of NFPA 921, by Chapter Subject

Chapter Title	2001 Edition	2004 Edition
Administration	Chapter 1	Chapter 1
Analyzing the Incident for Cause and Responsibility		Chapter 19
Appliances	Chapter 21	Chapter 24
Basic Fire Science	Chapter 3	Chapter 5
Basic Methodology	Chapter 2	Chapter 4
Bibliography	Appendix B	Annex B
Building Fuel Gas Systems	Chapter 7	Chapter 9
Building Systems	Chapter 5	Chapter 7
Cause Determination	Chapter 16	Chapter 18
Definitions	Chapter 1	Chapter 3
Documenting the Investigation	Chapter 13	Chapter 15
Electricity and Fire	Chapter 6	Chapter 8
Explanatory Material	Appendix A	Annex A
Explosions	Chapter 18	Chapter 21
Failure Analysis and Analytical Tools	Chapter 17	Chapter 20
Fire-Related Human Behavior	Chapter 8	Chapter 10
Fire and Explosion Deaths and Injuries	Chapter 20	Chapter 23

Continued

Chapter Title	2001 Edition	2004 Edition
Fire Patterns	Chapter 4	Chapter 6
Incendiary Fires	Chapter 19	Chapter 22
Legal Considerations	Chapter 9	Chapter 11
Management of Major Investigations	Chapter 24	Chapter 27
Mandatory References	Appendix C	Chapter 2
Motor Vehicle Fires	Chapter 22	Chapter 25
Origin Determination	Chapter 15	Chapter 17
Physical Evidence	Chapter 14	Chapter 16
Planning the Investigation	Chapter 12	Chapter 14
Referenced Publications	Appendix C	Annex C
Safety	Chapter 10	Chapter 12
Sources of Information	Chapter 11	Chapter 13
Wildfire Investigations	Chapter 23	Chapter 26

CHAPTER ORGANIZATION OF NFPA 921